HOW TO RAISE

RABBITS

EVERYTHING YOU NEED TO KNOW

SAMANTHA JOHNSON AND DANIEL JOHNSON

Voyageur Press

DEDICATION

To JEA, with love.

First published in 2008 by Voyageur Press, an imprint of MBI Publishing Company, 400 First Avenue North, Suite 400, Minneapolis, MN 55401 USA

Voyageur Press titles are also available at discounts in bulk quantity for industrial or sales-promotional use. For details write to Special Sales Manager at MBI Publishing Company, 400 First Avenue North, Suite 400, Minneapolis, MN 55401 USA.

To find out more about our books, visit us online at www.voyageurpress.com.

ISBN-13: 978-0-7603-4527-6

Library of Congress Cataloging-in-Publication Data

Johnson, Samantha.
  How to raise rabbits : everything you need to know / by Samantha Johnson and Daniel Johnson. -- Rev. and updated.
    p. cm.

Includes index.
ISBN 978-0-7603-4527-6 (softcover)
1. Rabbits. I. Johnson, Daniel, 1984- II. Title.
SF453.J53 2014
632'.6932--dc23
                                    2013021908

Editors: Amy Glaser, Elizabeth Noll
Design manager: James Kegley
Cover design: Carol Holtz
Interior design: Mandy Kimlinger
Layout: Danielle Smith

Photo credits
Front cover: Peredniankina/Shutterstock
Back cover: Daniel Johnson

Printed in China

10 9 8 7 6 5 4 3 2 1

CONTENTS

ACKNOWLEDGMENTS

It's always wonderful to have the opportunity to acknowledge all of the individuals who have provided helpful assistance on a book project, as it's our small way of saying "thank you." We would like to thank our editor, Amy Glaser, for her continued help, support, and enthusiasm, and especially the kind encouragement she passes along.

We would also like to thank the following individuals: Paulette Johnson for the endless hours spent editing photos for this book and for her help during photo shoots; Lorin Johnson for assistance with Chapters 3 and 10; J. Keeler Johnson for his photography contributions; Emily Johnson for help with Chapter 3 and assisting with rabbit wrangling; and Anna Johnson for proofreading and holding baby rabbits as needed. Also a big thank-you goes to Cheryl Sturtevant for checking in on our progress and generally keeping us hopping.

We also want to thank our circle of rabbit-enthusiast friends who contributed their knowledge and experience to us along the way during the writing of this book.

Thanks to Blake for always keeping things exciting, and to B. B. for seeing the best in Blake. Thanks also to Tessa, Sofie, Spring, Lottie, Bonnie, Woolimina, Peter, Pudge, and little Midge, without whom this book would not exist.

CHAPTER 1

SO, YOU WANT TO RAISE RABBITS

You are the only one who can answer that question. It may be as simple as knowing that you've always loved Rex rabbits and have always promised yourself that if you ever had the opportunity to raise rabbits, Rex would be your number-one choice. Or it might be a bit more complex. Perhaps you want to raise a popular breed so that you will have plenty of competition at the shows. Maybe you want to raise a breed that is recognized in a multitude of colors, or you might want to raise one specific color so that you can focus on achieving perfection in that particular shade. Perhaps you love lop-eared rabbits, or you don't. Maybe you love the idea of raising an Angora breed for the wool, or you can't imagine the grooming commitment. Perhaps you want to raise rabbits for meat, or you want to raise them for showing purposes.

A world of rabbits awaits you. Breeds, colors, shapes, sizes, and fur types are all a part of the fascination that keeps so many people interested in raising rabbits.

This Jersey Wooly is a Chestnut Agouti. Note the distinctive ring pattern where wind is blowing on her fur. The Jersey Wooly features Angora wool, a compact shape, and a small size.

This Tortoise-colored Holland Lop is an example of the compact shape. The Holland Lop is a small breed with normal fur.

Or perhaps you just don't know what's right for you. It is always good to start with the basics, which is why our first chapter is devoted to discussion of the 47 breeds recognized by the American Rabbit Breeders Association (ARBA), as well as information on the various sizes, shapes, fur types, and colors that you may encounter along the way. We'll also take a look at some of the rare-breed rabbits and discuss why you may want to consider raising them in your rabbitry. We'll also have a brief discussion of some of the newest breeds that have not yet received ARBA recognition. Let's get started.

NAVIGATING THE WORLD OF RABBITS

One of the first things that often surprises newcomers to rabbits is the vast array of breeds and varieties. With 47 breeds currently recognized by the ARBA, there are rabbits of every shape and size imaginable, and in every color, too. The scope of these breeds cannot be fully explored within the limitations of this chapter, I recommend my previous book, *The Field Guide to Rabbits*, to anyone wishing to gain knowledge of the details and history of each of the 47 rabbit breeds. In this chapter, however, we can certainly give each breed a quick overview.

If you're looking for the cute factor (and many of you probably are), then you can't go wrong with the Holland Lop, the American Fuzzy Lop, the Netherland Dwarf, or the Polish. With weights ranging from 2 to 4 pounds, these petite bunnies are inevitable crowd-pleasers and very popular with those who like to show. Entries for Holland Lops and Netherland Dwarfs usually outnumber most of the other breeds at shows.

If the lop-eared look catches your fancy but you would like something larger than a Holland Lop or an American Fuzzy Lop, you might want to consider one of the other lop-eared breeds, such as the Mini Lop, the English Lop, or the French Lop. Mini Lops are a mid-sized rabbit with an ideal weight of 6 pounds, while the English and French Lops are larger breeds, often 10 pounds or more.

If you like the fuzzy appearance of the American Fuzzy Lop but would prefer a rabbit breed with more traditional ears, then you might consider one of the other wooled breeds, such as the French Angora, the English Angora, the Satin Angora, the Giant Angora, or the Jersey Wooly. Of these breeds, the Jersey Wooly is the smallest, with the Giant Angora (you guessed it) the largest, and the French, English, and Satin breeds ranging in between. All of these breeds boast the gorgeous angora fur that makes these rabbits unique.

Some breeds are particularly noted for their distinctive color patterns. These breeds include the Dutch, the Californian, the Checkered Giant, the Hotot and Dwarf Hotot, the English Spot, the Harlequin, the Himalayan, and the Rhinelander. The Californian and Himalayan breeds have similar patterns that feature a creamy-white body accompanied by dark points (ears, nose, tail, feet) to create a very distinctive look. The Himalayan, however, exhibits an extremely different body type than the Californian, so don't worry about confusing the two breeds. The Hotot and Dwarf Hotot are entirely white with eyes encircled by a dark ring. The Harlequin is quite unique in that it is an unusual combination of calico coloring, accompanied by orange in the Japanese color variety and white in the Magpie color variety. Spots are the distinctive features of the Checkered Giant, the English Spot, and the Rhinelander, with each breed's standard differing slightly with regard to the placement of the spots. And finally, there is the one-of-a-kind Dutch, with its tuxedo-type coloring of a white chest and darker colored body.

The Tan breed is another small breed but it exhibits the full-arch shape. It features normal fur, and this particular rabbit is the Black Tan color variety.

These Rex rabbits exhibit the Lilac coloring. Rex rabbits are considered a medium-sized breed. They display the commercial shape and feature their distinctive Rex fur.

Two breeds have developed over time that are well known for their white coloring: the Florida White and the New Zealand White. Both breeds are commercial in type and are very popular. The New Zealand is also found in other colors, including Red, Broken, and Black, but White is the predominant color. On the opposite end of the color spectrum, you will find the Havana, which is similar in build to the Florida and New Zealand Whites, but is found in Black, Chocolate, Blue, and Broken.

Historically, chinchilla coloring was so popular that three ARBA-recognized breeds evolved with this specific coloring. These include the American Chinchilla, the Standard Chinchilla, and the Giant Chinchilla. Of these, the smallest is the Standard Chinchilla, with the American Chinchilla being a bit larger and the Giant Chinchilla larger still. The American Chinchilla was an extremely popular breed during the 1920s but has since slipped into the critical category on The Livestock Conservancy's list, meaning that very few American Chinchillas remain.

If you like the full arch variety of rabbit type, then you will definitely want to take a closer look at the Belgian Hare and the Britannia Petite. The Belgian Hare has the distinction of being the foundation of domestic rabbit interest in the United States during the rabbit boom at the turn of the twentieth century. The Britannia Petite is exactly as its name implies—petite—yet it is a very charming creature in spite of (or perhaps because of) its small size.

THE PLIGHT OF THE LIONHEAD

In recent years, the Lionhead breed has taken the United States by storm, appealing to a broad base of rabbit enthusiasts and increasing in popularity practically overnight. Yet it still has not reached breed status with the ARBA, despite extensive efforts by enthusiasts to reach the goal of having the Lionhead named as a recognized breed.

The Lionhead is a charming little breed, characterized by its trademark wooly mane. This distinctive trait is believed to have been the result of a genetic mutation that was first observed in the 1960s. The first examples were brought to the United States in 1998 and were approved at their first ARBA presentation in 2005. Subsequently, they failed their second and third presentations in 2006 and 2007, and though the Black, Tortoise, and Ruby-Eyed White varieties did pass their first presentation with a new COD holder in 2010, the varieties failed at their 2011 presentation. The Tortoise and Ruby-Eyed White varieties passed their second presentation at the 2012 convention, and are scheduled for their first attempt at a third presentation in 2013.

Despite the fact that they are currently without official ARBA recognition, the Lionhead remains extremely popular as a show rabbit, and more than 500 Lionheads were shown at the North American Lionhead Rabbit Club national exhibition in 2011. It is quite obvious that there is a great deal of admiration and support for these unique rabbits, and it is probably only a matter of time before they achieve their breed status.

Lionheads are quite unlike any other breed of rabbit. Their distinctive "mane" is quite captivating to many enthusiasts.

A bit of silver will brighten your day—and your rabbitry. The Silver, Silver Fox, Champagne d'Argent, and Silver Marten breeds are slightly different in coloring, but each features a very dramatic coat pattern that is extremely lovely. Of these, the Silver Marten is the most popular, while the Silver Fox is the largest and the most endangered. The Silver is on the threatened list, and it is also the smallest, weighing in at only 4 to 7 pounds. The Champagne d'Argent's name is French and translates to "the Silver rabbit from Champagne."

If typical rabbit fur doesn't interest you, there's always the possibility of getting started with Rex or Satin rabbits. Rex rabbits are sometimes known as velveteen rabbits because of their luxurious ⅝-inch coats, while Satin rabbits have coats that exhibit a luminous sheen. Both breeds conveniently come in smaller versions as well, the Mini Rex and the Mini Satin. All four breeds are enormously popular and well worth your consideration.

How about a mid-sized rabbit in an attractive color? You might consider the Tan or the newly recognized Thrianta. The Tan is a beautiful combination of dark fur with lighter points and is found in four varieties. The Thrianta is a deep orange-red and is rapidly gaining in popularity. If a larger rabbit catches your fancy, don't overlook all 13-plus pounds of the Flemish Giant.

Of course, we mustn't forget the more unusual breeds that we have not yet covered. These include the American, the American Sable, the Creme d'Argent, the Beveren, the Cinnamon, the Lilac, and the Palomino. Not all of these are on the Livestock Conservancy's list, but they are all considered unusual and are not always easy to find.

SIZES

From the most petite to the most massive, rabbit breeds span a size range from 2 pounds to 20. Many people end up choosing rabbits that are one of the sizes in the middle range, but there are enthusiasts and breeders of rabbits of all sizes who enjoy their breeds and wouldn't trade them for anything.

The Thrianta is one of the newest breeds recognized by the ARBA. It is a medium-sized rabbit with normal fur and a compact shape. Its standard color is a deep reddish-orange, which is shown here.

PETITE TO GIANT, WEIGHING IN ON SIZE

Although you needn't feel obligated to restrict yourself to only one breed, chances are that you're going to start out with only a few breeds in your rabbitry. As you consider all of the various breeds that are available for you to choose from, how do you determine the size that will be the most appropriate for your needs?

Contrary to the understanding of many rabbit newbies, the Giant breeds are not the breeds of choice if you are intending to raise meat rabbits. Their feed/meat conversion ratio is not nearly as good as that of the breeds that are slightly smaller, such as the New Zealand, the Satin, or the Champagne d'Argent. The Giant breeds are the ultimate in impressive show animals and are not as common as many of the smaller breeds, so they are often raised as breeding stock to maintain its presence.

The Flemish Giant is one of the largest rabbit breeds and tops the scales at more than 13 pounds.

The larger-end rabbits that are not Giant breeds are the most popular with breeders who are looking to raise rabbits for the purpose of meat. If this is your aim, then you will want to consider the breeds that have long track records of optimum results in this area.

Medium-sized rabbits, although sometimes used for meat, are more often used as pets and show (fancy) rabbits. Again, rabbit breeders often raise these medium-sized breeds as breeding stock to sell to other enthusiasts.

The small breeds, which include the adorable dwarf breeds, are the most popular as fancy rabbits and are very popular as pets. Obviously, the smaller breeds require less feed to maintain than a large or Giant breed, which can be a consideration if you're raising rabbits on a budget.

On the opposite end of the size spectrum, the Polish is one of the smallest rabbit breeds. It typically weighs less than 3½ pounds.

Let's take a quick look at the breeds in order by size, from the smallest to the largest.

Britannia Petite (up to 2½ pounds)
Netherland Dwarf (not more than 2½ pounds)
Dwarf Hotot (up to 3 pounds)
Jersey Wooly (not more than 3½ pounds)
Polish (not more than 3½ pounds)
Holland Lop (not more than 4 pounds)
American Fuzzy Lop (not more than 4 pounds)
Himalayan (2½ to 4½ pounds)
Mini Rex (3 to 4½ pounds)
Mini Satin (3¼ to 4¾ pounds)
Dutch (3½ to 5½ pounds)
Florida White (4 to 6 pounds)
Tan (4 to 6 pounds)
Thrianta (4 to 6 pounds)
Havana (4½ to 6½ pounds)
Mini Lop (4½ to 6½ pounds)
Silver (4 to 7 pounds)
English Angora (5 to 7½ pounds)
Standard Chinchilla (5 to 7½ pounds)
English Spot (5 to 8 pounds)
Lilac (5½ to 8 pounds)
Belgian Hare (6 to 9½ pounds)

Silver Marten (6 to 9½ pounds)
Satin Angora (6½ to 9½ pounds)
Harlequin (6½ to 9½ pounds)
Rhinelander (6½ to 10 pounds)
American Sable (7 to 10 pounds)
French Angora (7½ to 10½ pounds)
Rex (7½ to 10½ pounds)
Californian (8 to 10½ pounds)
Creme d'Argent (8 to 11 pounds)
Hotot (8 to 11 pounds)
Palomino (8 to 11 pounds)
Cinnamon (8½ to 11 pounds)
Satin (8½ to 11 pounds)
Beveren (8 to 12 pounds)
American (9 to 12 pounds)
Champagne d'Argent (9 to 12 pounds)
American Chinchilla (9 to 12 pounds)
New Zealand (9 to 12 pounds)
Silver Fox (9 to 12 pounds)
Giant Angora (9½ pounds and up)
English Lop (9 pounds or more)
French Lop (10½ pounds or more)
Checkered Giant (at least 11 pounds)
Giant Chinchilla (12 to 16 pounds)
Flemish Giant (13 pounds and over)

White markings are very important in many breeds, including Dutch (shown here), Rhinelander, English Spot, Dwarf Hotot, Himalayan, Hotot, and the Checkered Giant.

The Florida White is a small rabbit of compact shape. It features normal fur, and its standard color is, of course, White.

The smallest of the wooled breeds, the Jersey Wooly is a wooly bunny in a small package. Jersey Woolies typically weigh less than 3½ pounds. This doe is one of the larger examples of the breed and weighs over 3 pounds.

WHAT'S BEST "FUR" YOU?

Fur type is one more factor to consider when you are choosing a breed of rabbits to raise. Most breeds of rabbits possess what is known as normal fur, but there are several breeds with unique fur types that merit your consideration.

Satin fur is remarkable because of its luster and sheen. This is due to the hair shell's transparency. Satin fur is believed to have developed as the result of a mutation and was originally discovered in the Havana breed. Today the Satin is a breed in its own right, and the Mini Satin is a luminous smaller version.

Angora fur (or wool) is very distinctive, due in part to its length and in part to its woolen consistency. Six breeds exhibit angora fur: the French Angora, the English Angora, the Giant Angora, the American Fuzzy Lop, the Jersey Wooly, and the Satin Angora, which is particularly unique because of its interesting combination of satin and angora fur.

Rex fur, with its ideal length of ⅝ inch, is a particularly dense fur type, noted for its velvet-like softness and thickness. This density is due to the fact that the undercoat is the same length as the guard hairs, giving a very uniform appearance and extra softness. As is obvious from the name, rex fur is found in the Rex breed as well as the Mini Rex breed.

Of course, you might be satisfied with normal fur on your rabbits, in which case you have 37 rabbit breeds to choose from. One difference you will see within the various breeds with normal fur is that some breeds possess what is known as a "fly back" coat, meaning that the fur "flies back" into the original position after being brushed in the opposite direction. When a coat stays fluffed up, it is called a "roll back" coat. This is just one of the differences between fur types.

The Dutch is a small breed with the compact shape. It is found in six color varieties, including the Tortoise color shown here. The Dutch breed has normal fur.

This is a Standard Chinchilla rabbit, which exhibits Chinchilla coloring. The breed is a medium size with a compact shape and normal fur.

SHAPES

Circle, triangle, square, oval? Not quite. Rabbits are found in five shapes: full arch, semi-arch, compact, commercial, and cylindrical.

The most common shape of rabbit is the commercial shape, which is the type most found in breeds used for meat purposes. The breeds that fall into the commercial shape category include the following: the French Angora, the Giant Angora, the Satin Angora, the Champagne d'Argent, the Californian, the Cinnamon, the American Chinchilla, the Creme d'Argent, the French Lop, the Harlequin, the Hotot, the New Zealand, the Palomino, the Rex, the American Sable, the Satin, the Mini Satin, the Silver Fox, and the Silver Marten.

Next we have the compact shape, which is somewhat similar in looks to the commercial but distinguished by a shorter, more compact body, hence the name. Breeds that exhibit the compact shape include the American Fuzzy Lop, the English Angora, the Standard Chinchilla, the Dwarf Hotot, the Dutch, the Florida White, the Havana, the Holland Lop, the Jersey Wooly, the Lilac, the Mini Lop, the Mini Rex, the Netherland Dwarf, the Polish, the Silver, and the Thrianta.

Breeds that exhibit the full arch shape are notably taller than they are wide, accompanied by longer limbs. Full arch breeds include the Belgian Hare, the Britannia Petite, the Checkered Giant, the English Spot, the Rhinelander, and the Tan.

As a side note, it is interesting to observe that the Checkered Giant, the English Spot, and the Rhinelander all exhibit the same shape, as they are also somewhat similar in pattern and background.

Semi-arch breeds are also known as the mandolin type, due to their pear-shaped similarity to a mandolin. Only five breeds exhibit the semi-arch shape: the American, the Beveren, the English Lop, the Flemish Giant, and the Giant Chinchilla.

Last but not least, there is the cylindrical shape, which is exhibited by only one breed: the Himalayan. As implied by the name, the cylindrical shaped rabbits have lengthy, slender bodies.

COLORS

If you thought that there were a lot of rabbit breeds, just wait until you discover the array of colors in which rabbits can be found. It is truly a dazzling spectacle, and one that often surprises newcomers to rabbits.

The study of rabbit coat color genetics is a topic worthy of an entire book on its own, so we cannot fully detail each color and its genetic make-up within the limitations of this text. However, we can give you a brief rundown of the various breeds and colors to get you started on your way to understanding the correlation between breeds and the colors that they are found in.

Another color variety of Dutch is the Gray Dutch, which is shown here. The Gray Dutch color is actually an Agouti pattern, and this doe exhibits particularly nice ring color.

A delightful color, this Creme d'Argent depicts the standard color that the breed is known for. The breed is a large one, with normal fur and a commercial shape.

BREEDS RECOGNIZED IN ONLY ONE COLOR

These are the simplest to explain and understand, so it seems logical that we should start with the breeds that are found in only one color, their "standard" color recognized by the ARBA.

American Sable

As suggested by the name, this breed's standard color is Sable, which is a deep, dark brown with lighter areas on the back, shoulders, and inside the ears.

Giant Angora

At the time of this writing, the Giant Angora is recognized only in the Ruby-Eyed White color, a very striking combination of a pure white coat accompanied by ruby-red eyes.

Belgian Hare

The standard color for Belgian Hares is a rich reddish color, although the individual shades may vary from Tan to Chestnut. This coloring is accompanied by ticking.

Californian

This distinctively colored breed has lent its name to its color, which is White with Gray-to-Black points (ears, nose, feet, and tail). Many other breeds may exhibit Californian coloring, but the Californian breed is found only in this color.

Champagne d'Argent

This French breed exhibits the Silver color, a combination of Blue, Black, and White hairs with darker patches on the face.

American Chinchilla, Standard Chinchilla, and Giant Chinchilla

These three breeds have something very important in common, their distinctive Chinchilla coloring. This unique color is best described as having hairs that are Blue, then Gray, then White, and tipped with Black.

Cinnamon

This one is a bit confusing. The Cinnamon breed features Cinnamon coloring, which is rust- or cinnamon-colored with gray ticking. However, some other breeds (notably the Angora breeds) also recognize a color known as Cinnamon, but this is a different genetic combination and is sometimes called Chocolate Agouti or Amber.

Creme d'Argent

Another unique color found only in this beautiful breed, the Creme d'Argent is always a shade of orange-silver coloring with a creamy appearance, hence the name.

Florida White

What color could the Florida White possibly be? Yes, you guessed it, these rabbits are White. As is obvious from the name, this is the only color that is recognized in this breed.

Hotot and Dwarf Hotot

At first glance, these two breeds may look like just another white rabbit. But upon closer examination, you can't help but notice the distinctive dark band encircling their eyes. In the Hotot breed, the band is always Black, but the Dwarf Hotot can have a band that is Black, Chocolate, or Blue, although the Blue variety is not yet officially accepted by the ARBA.

Lilac

A little bit pink and a little bit gray, the Lilac breed and color is quite unlike any other. The beautiful shading of gray on pale brown gives the effect of lilac coloring. As with the Cinnamon, the term "Lilac" can be slightly confusing because there is a breed called Lilac, yet the color of Lilac is found in several additional breeds, including the American Fuzzy Lop, the English Angora, the French Angora, the Satin Angora, the Jersey Wooly, the English Lop, the French Lop, the Holland Lop, the Mini Lop, the Netherland Dwarf, the Rex, and the Mini Rex.

Silver Fox

The only variety of the Silver Fox coloring that is currently recognized is Black, but it must be noted that it is not black in the typical sense, but rather a Black that is accompanied by Silver ticking throughout the coat.

Thrianta

And finally we have the Thrianta, which is particularly noted for its striking coloring, a deep orange-red.

BREEDS RECOGNIZED IN TWO COLORS
American

This breed's two recognized varieties are Blue and White, although there is the possibility of a Red color being developed for future recognition with

The Palomino is another large breed that also has the commercial shape and normal fur. This breed is found in two colors: Lynx and Golden (shown here).

the ARBA, and it would certainly be rather fitting for the breed of American rabbit to be recognized in Red, White, and Blue.

Checkered Giant
This breed of spotted rabbit has a White base coat accompanied by markings in either Blue or Black. These markings must follow a specific pattern as outlined in the *Standard of Perfection*.

Harlequin
The calico coloring found in this breed is entirely its own, although some breeders are attempting to produce a Harlequin-colored Dutch, which is as yet unrecognized by the ARBA. The Harlequin breed is found in two color varieties: Japanese and Magpie, with the former comprising Black, Blue, Chocolate, and Lilac accompanied by orange. The latter comprises Black, Blue, Chocolate, and Lilac accompanied by White.

Palomino
The Palomino breed (and color) is a golden tan color—very similar to a palomino horse—and there

are also two recognized varieties of the Palomino: Lynx and Golden.

BREEDS RECOGNIZED IN MULTIPLE COLORS
The following breeds have received ARBA recognition in multiple colors:

American Fuzzy Lop
There are nineteen colors currently recognized for American Fuzzy Lops, including Chinchilla, Chestnut, Lynx, Opal, Squirrel, Pointed White, Black, Blue, Blue-Eyed White, Chocolate, Lilac, Ruby-Eyed White, Sable Point, Siamese Sable, Siamese Smoke Pearl, Tortoiseshell, Fawn, Orange, and Broken.

English Angora, French Angora, Satin Angora
These three breeds are all recognized in the same colors, which include Pointed White (Black, Blue, Chocolate, and Lilac varieties), Blue-Eyed White, Ruby-Eyed White, Chinchilla, Chocolate Chinchilla, Lilac Chinchilla, Squirrel, Chestnut, Chocolate Agouti, Copper, Lynx, Opal, Broken, Black, Blue, Chocolate, Lilac, Pearl, Sable, Seal, Smoke

19

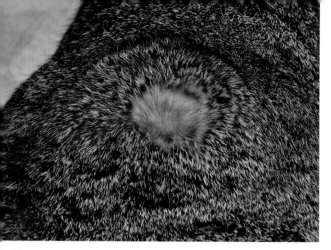

If you blow into the fur on an Agouti-colored rabbit, you will be able to see what is called the "ring color." Note the gradual shading of color from the hair tips all the way down to the hair base.

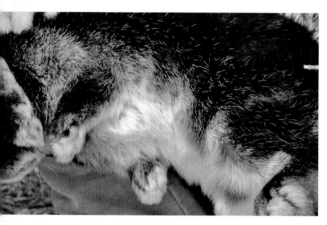

The otter pattern is found in several different colors, but all share the same characteristics of lighter shading under the belly.

This is a close-up depicting the Chinchilla coat pattern. Note how the gray hairs are lighter at the base and darker at the tips.

Pearl, Tortoiseshell, Blue Tortoiseshell, Chocolate Tortoiseshell, Lilac Tortoiseshell, Blue Steel, Chocolate Steel, Lilac Steel, Steel, Cream, Fawn, and Red.

Beveren
In sharp contrast to the Angora breeds listed above, the Beveren is found in only three colors: White, Black, and Blue.

Britannia Petite
This tiny breed is found in six colors: Black, Black Otter, Chestnut Agouti, Ruby-Eyed White, Broken, and Sable Marten.

Dutch
Although the pattern of white markings is extremely distinctive and important, the coloring of the dark portions of the Dutch breed can be any of these varieties: Black, Blue, Chocolate, Gray, Steel, and Tortoise. Three additional shades that have not yet achieved recognition with the ARBA are the Harlequin Dutch, the Chinchilla Dutch, and the Gold Dutch. All three varieties are strikingly beautiful.

English Spot
As with the Dutch breed, the color varieties of the English Spot refer to the darker spots and markings that are accompanied by white. The recognized colors for English Spots include Black, Blue, Chocolate, Gold, Gray, Lilac, and Tortoise.

Flemish Giant
This large breed is found in seven colors: Black, Blue, Fawn, Light Gray, Sandy, Steel Gray, and White.

Havana
Like the New Zealand, the Havana is found in three colors, but instead of White, Red, and Black, the Havana is recognized in Chocolate, Blue, and Black. An additional fourth color, Broken, achieved official recognition as of the 2007 ARBA National Convention.

Himalayan
There are four varieties of the Himalayan coloring, which is a White body accompanied by colored points on the feet, tail, nose, and ears. The colored points can be Black, Blue, Chocolate, or Lilac.

Jersey Wooly

Another breed that is found in a multitude of colors, the Jersey Wooly is currently recognized in these varieties: Chestnut, Chinchilla, Opal, Squirrel, Pointed White (Black or Blue varieties), Broken, Black, Blue, Blue-Eyed White, Chocolate, Lilac, Ruby-Eyed White, Blue Tortoiseshell, Sable Point, Seal, Siamese Sable, Smoke Pearl, Tortoiseshell, Black Otter, Blue Otter, Sable Marten, Silver Marten (Black, Blue, Chocolate, or Lilac varieties), and Smoke Pearl Marten.

English Lop, French Lop, and Mini Lop

The English, French, and Mini Lops share the same list of colors, which include the following: Chinchilla, Chestnut Agouti, Lynx, Opal, Broken, Tri-colored, Black, Blue, Chocolate, Lilac, Ruby-Eyed White, Blue-Eyed White, Frosted Pearl (Blue, Black, Chocolate, or Lilac varieties), Sable, Sable Point, Seal, Smoke Pearl, Tortoise (Black, Blue, Chocolate, or Lilac varieties), Silver/Silver Fox (Black, Blue, Brown, or Fawn varieties), Steel, Cream, Fawn, Orange, and Red.

This Rex rabbit exhibits Broken Chinchilla coloring. A rabbit with Broken coloring must have a Broken-colored parent.

A slightly smaller version of the Rex rabbit, this Mini Rex also features Rex-type fur. The Mini Rex features the compact shape and is a small breed. This example displays the Blue color.

Holland Lop

Like the American Fuzzy Lop, the Holland Lop is recognized in a list of colors that differs from those of the English, French, or Mini Lop breeds. Holland Lops are recognized in these colors: Chestnut Agouti, Chocolate Agouti, Chinchilla, Chocolate Chinchilla, Lynx, Opal, Squirrel, Broken, Tri-colored, Pointed White (Black, Blue, Chocolate, or Lilac varieties), Black, Blue, Chocolate, Lilac, Blue-Eyed White, Ruby-Eyed White, Sable Point, Siamese Sable, Seal, Smoke Pearl, Tortoise (Black, Blue, Chocolate, or Lilac varieties), Otter (Black, Blue, Chocolate, and Lilac varieties), Steel (Black, Blue, Chocolate, or Lilac varieties), Cream, Fawn, Frosty, Orange, and Red.

Mini Satin

For a breed that only achieved its ARBA recognition in 2006, the Mini Satin has made big strides in a short time. Several colors have been officially accepted as recognized varieties: Ruby-Eyed White, Chinchilla, Opal, Red, Black, Chocolate, Otter, and Siamese. Additional colors are currently in development.

This is a Silver Marten with Blue coloring. The Silver Marten breed is a medium-sized breed with the commercial shape and normal fur.

The Harlequin is a medium breed of the commercial shape. This particular example is a Blue Magpie Harlequin and has normal fur.

Netherland Dwarf

Twenty-six colors are currently recognized in the Netherland Dwarf breed: Black, Blue, Chocolate, Lilac, Blue-Eyed White, Ruby-Eyed White, Sable Point, Siamese Sable, Siamese Smoke Pearl, Chestnut, Chinchilla, Lynx, Opal, Squirrel, Otter (Black, Blue, Chocolate, or Lilac varieties), Sable Marten, Silver Marten (Black, Blue, Chocolate, or Lilac varieties), Smoke Pearl Marten, Tan (Black, Blue, Chocolate, or Lilac varieties), Fawn, Himalayan (Black, Blue, Chocolate, or Lilac varieties), Orange, Steel, Broken, and Tortoiseshell (Black and Blue varieties).

New Zealand

The New Zealand is found in four colors: White, Red, Black, and the newly recognized Broken.

Polish

Unlike some of the other petite breeds that are found in a multitude of colors, the Polish is recognized only in Black, Blue, Broken, Chocolate, Blue-Eyed White, and Ruby-Eyed White.

Rex

The Rex is found in many of the standard colors that are often seen in rabbit breeds, including Black, Black Otter, Broken, Californian, Chinchilla, Chocolate, Lilac, Lynx, Opal, Red, Sable, Seal, and Ruby-Eyed White. However, the Rex is also found in its own unique color, known as Castor, which is a deep, dark brown.

Mini Rex

Like the Rex, the Mini Rex is also recognized in the Castor color variety. Other Mini Rex colors include Black, Blue, Blue-Eyed White, Broken, Tri-color, Chinchilla, Chocolate, Himalayan (Black or Blue varieties), Lilac, Lynx, Opal, Red, Seal, Tortoise, Ruby-Eyed White, Silver Marten, and Smoke Pearl.

Rhinelander

Predominantly White, the Rhinelander is accompanied by Black-and-Tan spots. A second variety, newly recognized by the ARBA, has Blue-and-Fawn spots.

Satin

The remarkable fur type of the Satin rabbit is simply beautiful in any color. Recognized colors in the Satin breed include Black, Blue, Broken, Californian, Chinchilla, Chocolate, Copper, Otter, Red, Siamese, and Ruby-Eyed White.

Another medium-sized breed, the Rhinelander exhibits the full-arch shape and normal fur. This Rhinelander is White with Black-and-Tan markings.

This French Lop is notable for its distinctive lopped ears. The French Lop is a large breed of the commercial shape and features normal fur. This rabbit's color is Chestnut Agouti.

Here is another rabbit with Chestnut Agouti coloring, but in a much smaller version. This petite Netherland Dwarf is a small rabbit of compact shape with normal fur.

A medium-sized rabbit of compact shape, the Mini Lop has normal fur, and this particular example is Blue Opal.

This Rex rabbit features Chocolate coloring, which is a deep rich brown. This color is found in several breeds, including all of the Lop varieties, as well as three of the Angora breeds.

Silver
The distinctive coat pattern of the Silver breed is currently recognized in three varieties: Black, Brown, and Fawn.

Silver Marten
The Silver Marten breed is found in four shades: Black, Blue, Chocolate, and Sable. These shades are accompanied by silver markings and hair tips.

Tan
Strikingly impressive with their unusual coat pattern, the Tan breed has four varieties: Black, Blue, Chocolate, and Lilac. Each of these varieties is accompanied by tan/orange markings.

RARE BREED RABBITS
With so many options available to you, why would anyone need to look beyond the typical Rex or Netherland Dwarf or Holland Lop to find what they are looking for? In many cases, rabbit enthusiasts are quite content with the usual array of breeds and feel no need to delve further into the list of rabbit breeds.

However, people are becoming increasingly intrigued by the idea of raising some of the rarest rabbit breeds in an effort to help preserve and perpetuate these uncommon (and sometimes endangered) breeds. Raising rare breeds can be a challenge in many ways, but many people feel that the benefits far outweigh any difficulties.

The English Angora is a medium-sized breed with the compact shape, as well as with Angora fur. This example is Fawn colored.

A full-arch breed, the Britannia Petite is small and features normal fur. This particular Britannia Petite displays Black Otter coloring.

A giant-sized breed, the Flemish Giant is a semi-arch shape with normal fur. This rabbit is Light Gray.

The Satin breed is noted for its remarkable satin fur. The breed has the commercial shape and is a large breed. This rabbit features Broken Red coloring.

Finding foundation stock for rare breeds can be one of the most challenging areas for newcomers. Because the number of animals is so limited on a national basis, finding foundation stock in your general proximity can be very difficult. It can also be difficult to locate stock that is of sufficient quality for breeding. In many cases, rabbits must be shipped cross-country in order to get your breeding program started on the right foot. Some breeds are rare because of other factors, such as a high rate of litter mortality, or because the does in the breed are not noted for being good mothers. In other cases, the rarity is compounded by the fact that producing an example that meets the breed's standard of perfection is a difficult task. For instance, the Blanc de Hotot must exhibit the proper pattern of markings in order to meet the standard. In every litter, there is the chance of producing kits that do not meet the proper criteria for markings. This makes them unfit for showing or registration purposes. The same is true with the critically endangered American breed. Only the Blue and White colors are recognized, yet many other colors appear in American litters on a regular basis, which cuts down the number of kits with acceptable coloring that are born each year.

Of course there are the benefits. There is something exceptionally rewarding about helping preserve a breed that might otherwise slip away into oblivion, utterly eclipsed by other more popular breeds. There is something very gratifying about knowing that your rabbitry is home to one of the rarest and most unusual breeds of rabbits in the United States. By producing litters, you are helping to increase the population and ultimately to save the breed from extinction. From a business perspective, you may find that you have a more demanding market for the rarer breeds simply because there are so few being produced. Many rare-breed rabbitries have waiting lists of buyers who pay in advance for the results of future litters because they sell out so quickly.

LEFT: The only breed that exhibits the cylindrical shape is the Himalayan, which also features its distinctive Himalayan coloring. This particular rabbit has black points. It is a small breed with normal fur.

BELOW: The New Zealand breed is found in White, Red, Black and Broken, but the White version shown here is the most commonly found. The New Zealand breed has the commercial shape and is a large breed with normal fur.

If raising a rare breed interests you, let's take a look at what constitutes a rare breed. The Livestock Conservancy recognizes 11 rabbit breeds on their 2012 Conservation Priority List. Of these, the American Chinchilla is recognized as the rarest and is listed as critical on the 2012 list. A critical rating means that it is estimated that there are less than 500 rabbits of this breed remaining in the world. The next category is threatened, with fewer than 1,000 estimated worldwide.

On the threatened list are the American, the Belgian Hare, the Blanc de Hotot, the Silver, and the Silver Fox. It seems particularly ironic that the American Chinchilla and the Belgian Hare have found places on these lists, as both breeds were enormously popular during the first 20 years of the twentieth century. It's nearly impossible to comprehend that these two breeds have declined to such an extent as to be listed as critical and threatened.

This Netherland Dwarf features Siamese Sable coloring, a color that is only found in three other breeds: the American Fuzzy Lop, the Jersey Wooly, and the Holland Lop.

Like the Rhinelander, the Checkered Giant is a full-arch breed, but it is a Giant breed. The Checkered Giant has normal fur and has Broken Blue coloring.

A small breed of compact shape, the Polish breed has normal fur. This example is Black.

29

Continuing on, the Livestock Conservancy recognizes four breeds on its watch list, which generally means that there are less than 2,000 remaining worldwide. The breeds on this list include the Beveren, the Giant Chinchilla, the Lilac, and the Rhinelander. An additional breed, the Creme d'Argent, is listed as recovering.

Although they are not listed on the Conservation Priority List, there are several other rabbit breeds that are very unusual in the United States. These include the Velveteen Lop, a breed that is working toward recognition by the ARBA but has not yet achieved its fully recognized status; the Perlfee, a newly introduced breed from Germany; the Astrex, an unusual type of Rex rabbit that is nearly extinct, with only a few lines remaining worldwide; and many others.

So, how do you get started with a rare breed? It isn't easy, and you can't expect to establish a breeding program as easily as you might if you were starting with a more common breed. You will have to do your homework. The Livestock Conservancy's member directory is a good place to start, although many rare-breed rabbit breeders are not members and are not listed in the directory. An Internet search might result in some good leads that could put you on your way to locating a reputable breeder. Ask around and check with other rabbit enthusiasts who might be able to put you in touch with a breeder. Don't expect to find what you're looking

HANDS-ON EDUCATION

You may have created a short list of breeds that have caught your fancy and have begun to wonder if you would like to raise them yourself. Before you commit to any rabbit breed, you will want to take some time to be around them in person. Visit a breeder that specializes in your chosen breed and ask to see many different examples. Get a real feel for their size and appearance. Pictures and descriptions in books or on the Internet can be a little deceiving, so unless you've had the opportunity to evaluate a breed in person, you might be surprised by how different it is from your mental image. It's hard to imagine just how tiny a Netherland Dwarf is until you've held one in your hands. Likewise, it's hard to visualize the huge size of a Flemish Giant unless you've visited with one in person. Visiting a rabbit show can be a great way to view a multitude of breeds in one location and get a feel for their relative sizes and appearances.

MUST-DO'S FOR THE NEW RABBIT OWNER

If there's one piece of advice that we can't repeat enough, it is that you should join the ARBA. It is our opinion that every rabbit owner (and especially every rabbit breeder) should be an ARBA member. The annual membership fee is nominal, and the benefits are worth every penny. You will receive six issues per year of the ARBA's magazine, *Domestic Rabbits*, as well as the new-member packet, which contains a copy of the yearbook and the ARBA's guide to successful rabbit raising. Being an ARBA member also allows you to register your rabbitry name, allows your rabbits to be eligible for points ("legs"), and to be granded at shows.

In any case, you will want to join the ARBA simply to receive *Domestic Rabbits* every other month. It is jam-packed with reports on specialty breed clubs, breed history, and shows, plus it contains up-to-the-minute information on breeds and varieties that are currently in development. Besides all of this, there are advertisements to put you in touch with breeders, equipment suppliers, feed companies, and software suppliers, as well as information on the ARBA's annual national convention.

for in your backyard—or even in your state. It's possible that you might be fortunate enough to live near a wonderful producer of rare-breed rabbits, but chances are that you will have to ship your rabbits in from long-distance in order to obtain the breed (and the quality) that you are seeking.

There is one resource that receives my highest recommendation for anyone wishing to get started with rare breeds, and that is the Rare Breed Rabbit Internet discussion group. The website address is http://pets.groups.yahoo.com/group/rarebreedrabbits/. This is a wonderful group of dedicated enthusiasts with years of experience and knowledge. They offer listings of rare breed rabbits for sale, discussions of various breeds and ways to promote them, and much more.

The Silver Fox is one of the rarest rabbit breeds in North America. It is currently listed as threatened on The Livestock Conservancy Conservation Priority List.

CHAPTER 2

ESTABLISHING YOUR RABBITRY

Now that you have acquainted yourself with a brief overview of rabbits in general and the 47 ARBA-sanctioned breeds in particular, it's time to focus on the task that lies before you: establishing your rabbitry. In subsequent chapters we will fully discuss the basics of housing, feeding, and health care for your rabbits, but for the moment, let's concentrate on the actual basics of choosing your foundation stock and purchasing the first rabbits for your breeding program. Understandably, this step is a mixture of excitement and hesitation. You're eager to begin raising rabbits and you can't wait to get started, but you're a little uncertain about how to make the right choices. In this chapter, we'll explain the different options you have for purchasing rabbits and how to select your stock once you've located a breeder.

When establishing your rabbitry, it's always wise to start off with the finest foundation stock that you can locate and afford. Top-quality rabbits will help you establish your breeding program in a quicker fashion than if you started off with stock of poorer quality.

These two rabbits illustrate distinct differences in physical appearance. Despite being similar in size and both exhibiting Tortoise coloring, these two are very different. The Holland Lop on the left displays lop ears, while the Dutch doe on the right exhibits the distinctive pattern of white markings that the Dutch breed is noted for.

ONE BREED . . . OR MORE?

When you're just getting started, it may seem logical to choose one breed of rabbit and focus on it, investing all of your time and effort into selecting the best possible breeding stock, starting your rabbitry, and producing quality examples of the breed. However, the truth is that people rarely start with just one breed.

Due to the fact that there are 47 ARBA-recognized breeds, there is definitely a tempting assortment to choose from. Add to this the handful of additional breeds that are in various stages of development, with breeders needed to help establish them, and you'll understand why it's hard to focus on just one.

Space and time limitations will undoubtedly curtail you from dabbling in all 47 breeds or even a dozen, but it's perfectly acceptable to choose three or four breeds to work with. As your rabbitry progresses, you will probably discover that you've developed a particular preference for some breeds, and you'll probably find that others do not thrill you as much. This understanding will help you decide which breeds to continue pursuing in the future and which ones you might not care to work with anymore.

There really is no right or wrong choice when it comes to having one breed or multiple breeds. Some breeders prefer to work with only one breed and become known for their top-quality French Lops or their wonderful Netherland Dwarfs. Other breeders might specialize in wooly breeds and raise English Angoras, French Angoras, Satin Angoras, Giant Angoras, and Jersey Woolies in their rabbitry. Many breeders nowadays are focusing solely on working with the rare heritage breeds, such as Americans, American Chinchillas, Silver Foxes, and Beverens. Then again, you will also find breeders who raise a range of types and sizes—Polish and Flemish Giants, Mini Lops, and New Zealand Whites—with seemingly no rhyme or reason to their choices. All of these options are good ones, and it's really a matter of deciding what works best for you and your rabbitry and then working hard at promoting your chosen breed or breeds.

There are many questions that you will have to answer before you settle upon choosing a breed. It's even possible that you will choose a breed and work with it for a time before you realize it really doesn't suit your purposes. Some people raise the same breed of rabbit for decades, while others "hop" around, so to speak, from breed to breed, dabbling in each for a time before proceeding on to the next one. Until you've been raising rabbits for a while, you really will have no way of knowing which type of breeder you are.

CHOOSING FOUNDATION STOCK

Decision-making can seem a little daunting when you first start off with the intention of raising rabbits. Your initial choices can have a profound and long-lasting effect on the success or failure of your venture, so you will obviously want to make sure that you choose carefully when selecting breeding stock.

It's easy to buy rabbits. All you have to do is head down to your nearest pet shop, and you'll undoubtedly find plenty to choose from. Delightful as they may be, they are probably only "pet quality" rather than "breeding quality" rabbits. Beginning breeders often make the common mistake of rushing into their purchases. Once the decision to raise rabbits has been made, they want the rabbits immediately. They have made plans and they want to put the plans into action now. Hasty purchases are often made in an effort to get their hands on any furry animal with long ears and an affinity for carrots, regardless of quality, type, or pedigree.

That's why it's important to remember that having 10 mediocre rabbits is not better than having 3 top-quality rabbits. It is an oft-quoted

saying, but it bears repeating: focus on *quality*, not *quantity*. If you channel your energy into striving to produce rabbits of the highest quality, your breeding program will be so much the better because of it. This is not to say that a person cannot produce quality and quantity, as that can and does happen.

A trio, like this trio of Champagne d'Argent rabbits, is a very good way to begin your breeding program. By purchasing a buck and two does, you can begin producing litters without initially going in over your head.

With many breeds, such as the Dutch (pictured here), the English Spot, or the Hotot, the pattern of markings is extremely important, so remember to evaluate the markings of any rabbit that you're considering for your breeding program.

The ARBA's *Standard of Perfection* must always be your guide, and this is particularly vital when you are selecting rabbits to purchase. Don't hesitate to ask a breeder if you can examine a couple of rabbits in comparison with one another.

This Rex is a beautiful example of a beautiful breed. If you spend a lot of time looking at high-quality rabbits, it will help educate your eye so that you can easily spot a good example.

But for a beginner breeder just starting out, the tendency can be to get started quickly with "a bunch of rabbits" in order to establish a viable rabbitry without really considering the long-term effects of those choices.

Another aspect to consider is how quickly your number of rabbits can increase. It's not necessary to start out with a dozen rabbits. A *trio* (one buck, two does) is an ample number to get you started on the road to raising rabbits. Their subsequent litters will allow you to quickly increase your number of rabbits with only selective additions of purchased rabbits later on.

Let's say that you have set out to find the trio of rabbits that will establish your rabbitry and jump-start your dreams of raising rabbits. What characteristics should you look for? What's vital and what's not? How can you choose the best rabbits, and where do you start looking?

Although it may be an obvious statement, the most important qualification is that the rabbits you purchase should be healthy. This means that they should be free of any type of congenital defects or contagious diseases. You will want to purchase only rabbits that are absolutely and entirely the picture of perfect health. This means that they should be free of any type of disqualifying defect, such as malocclusion of the teeth, blindness, mismatched toenails, or a torn ear. Never compromise in these areas, as they are vitally important to the future welfare of your rabbitry, and you want your foundation stock to be as nearly perfect as you can find.

This is a Harlequin rabbit of less-than-stellar quality. This might make a wonderful pet rabbit, but it's not the type of rabbit that you would want to use as a foundation for your breeding program.

This is a higher-quality example of a Harlequin rabbit and would make a better choice for a breeding program than the rabbit in the photo above. If you can't distinguish or decide between the rabbits that you're looking to buy, then you might want to ask the advice of a knowledgeable rabbit breeder.

Before purchasing any rabbits, it is highly recommended that you thoroughly acquaint yourself with the ARBA's *Standard of Perfection* for your chosen breed. Without a working knowledge of your breed's standard, you will not be able to make the best choices in securing your foundation stock. Always consider all of the information listed in the *Standard of Perfection* before you make any selections to purchase. People occasionally dismiss the *Standard of Perfection* as being nothing but the criteria for which rabbits are judged at shows. On the contrary, it's important to remember that as a breeder, it is your responsibility to produce rabbits that meet the standard of perfection as completely as you possibly can, whether your rabbits ever set foot on a show table or not. Therefore, your foundation stock should always be excellent examples of their breed, which will in turn put your fledgling rabbitry on its way toward producing litters with good type and quality.

It is also very helpful to view as many examples of the breed as you can. It is sometimes hard to visualize exactly what a "narrow shoulder" or a "weak hind leg" looks like, unless you have had the experience of evaluating numerous rabbits against each other and against the standard. The quickest way to train your eye for quality is to examine outstanding examples of the breed. If you can look at the very best and imprint the characteristics in your mind, then you will find that the substandard rabbits quickly make themselves quite evident in poor comparison. It's even better if you can hold the very best in your hands to see what a good rabbit looks like up close. A great way to evaluate many rabbits at the same time is to spectate at rabbit shows. The judge's comments are extremely enlightening, particularly to a rabbit enthusiast who is just starting out with his or her breeding program.

In addition to breed type, markings, and color, your breeding stock must also be free of physical defects, so always take the time to check the rabbit's teeth and toenails, among other areas.

When you're starting out, you might want to purchase rabbits that are old enough to begin breeding, so as to jump-start your breeding program. On the other hand, if you're not in a particular hurry to get started, you might want to purchase young rabbits and wait for them to grow up. Don't worry, it doesn't take long.

CONSIDERATIONS WHEN CHOOSING FOUNDATION STOCK

Throughout the history of domestic rabbit breeding, unexpected colors have cropped up in litters where they were not anticipated. They became known as "sports," a term specifically applied to purebred rabbits that exhibited colors that were not of a recognized variety. For instance, the Lilac breed was developed from sports that appeared in the Havana breed, which was only recognized in Black, Blue, and Chocolate. The Lilac color would appear when the dilute gene of the Chocolates would combine with the Blue to result in Lilac. These Lilac-colored sports eventually became what is known today as the Lilac rabbit.

What does this mean for your rabbitry? When you're first getting started, it's important to be fully acquainted with the recognized color or colors in your breed so that you can focus on purchasing foundation stock that exhibits the proper coloring. This is not to say that you won't get the occasional sport of your own in your litters, but by carefully selecting your breeding stock, you can help maximize your chances of producing rabbits in recognized colors.

It also helps if you have a solid understanding of color genetics in rabbits. If you know the basics of the color combinations and understand the results of crossing Rabbit A with Rabbit B, this will also help reduce the chances of producing rabbits in unrecognized colors.

Lilac coloring, as shown here, is found in the following breeds: American Fuzzy Lop, English Angora, French Angora, Satin Angora, Jersey Wooly, Lilac, English Lop, French Lop, Holland Lop, Mini Lop, Netherland Dwarf, Rex, and Mini Rex.

Another important area of consideration is age. There are a few different ways to look at the subject of age, and your personal preference will certainly dictate the route you take. On one hand, it's never unwise to purchase young rabbits. Because rabbits mature so quickly and are able to reproduce at a young age, you can purchase newly weaned kits and still be in possession of your first litters within a year's time. Unlike many other livestock breeds, there is very little waiting around for rabbits to grow up. Young rabbits have all of their years of productivity ahead of them to maximize your potential for litters. Younger rabbits also are less likely to possess health issues that may interfere with reproduction.

If you want to raise rabbits in a multitude of colors, then you might want to consider one of the breeds that is found in a wide variety of shades, such as the American Fuzzy Lops seen here.

On the other hand, purchasing older rabbits has its benefits as well. Older rabbits (as long as they are still in their prime years of reproductive health) have the benefit of being proven. This takes away a lot of the guesswork in purchasing a rabbit. If you can evaluate the offspring of a buck or doe, you can quickly decide whether or not they are producing the type and quality you're seeking. Similarly, older rabbits may have show experience or even show titles to their credit, which can make them a more valuable asset to your breeding program, especially as you get started. Of course, the downside to purchasing older rabbits is that they have less remaining time of prime production, which can decrease the number of litters that you may expect from them, but a well-proven rabbit can go a long way toward getting your rabbitry off on the right foot. These are decisions that you will have to carefully weigh in your mind before choosing your foundation stock.

One other point to ponder is that you may or may not have the option to be choosy about the age of the rabbits you're looking to purchase, at least not if you're also being selective about quality and type. You can easily find quality rabbits of popular breeds available for sale at all times and at all ages, making it very easy to get started if you're interested in raising Netherland Dwarf or Holland Lop rabbits. But if your fancy is captured by one of the rarer breeds, such as the American Chinchilla or the Silver Fox, you may have a harder time locating rabbits for sale, let alone rabbits of a particular age. In this case, you may have to content yourself with whatever you can find that is available and meets your other criteria for purchase. You can always compromise on age and purchase a rabbit that is a bit younger or older than your preferred age, but you will still want to always remain firm in your standards in other areas, such as health, disposition, and breed type. These are areas in which you cannot compromise.

Price may be a consideration, and it's possible that a junior rabbit may be a bit more affordable than a senior rabbit that already possesses a show record and a production record. On the other hand, proven breeding stock that has reached "middle age" may be more reasonably priced than a quality youngster. This is because older rabbits are usually past being shown, so you would be purchasing the rabbit as a brood rabbit only. This is not to say that they aren't of good quality, just that they are showing a bit of age and their reproductive time may be running short. In these cases, you can sometimes obtain very nice stock at a fraction of the cost of buying youngsters of similar quality.

You may also find that while physical characteristics can be easy to spot, it can be harder to pinpoint that elusive area of "type," those intrinsic qualities and subtle characteristics that mark the difference between an average rabbit and a superior one. Again, this is simply a matter of education and training your eye, and unfortunately it doesn't happen overnight. But when it does happen and you can look with confidence at a class full of junior bucks and know which ones are the best and why, it is an incredibly rewarding achievement.

Although you may place color far down on your list of characteristics to consider when shopping, a quick look at your breed's description in the *Standard of Perfection* can illustrate the importance of this seemingly trivial area. In many rabbit breeds, color is a very important part of the breed's description, and it will be something that you will select for when raising litters. A breed's ideal coloring is always an area to consider, since you will want to produce rabbits that exhibit proper coloring, particularly if you intend to show or wish to sell show prospects.

Similarly, for certain breeds, such as the English Spot or the Rhinelander or the Dutch, the pattern of markings is also extremely important. Any deviation from the established pattern (as outlined in the *Standard of Perfection*) is viewed as imperfect, and therefore you will always want to select foundation stock that exhibits the ideal pattern as closely as possible. By establishing yourself with breeding stock that meets the standard, you can increase your odds of producing correctly-marked babies. Obviously, color genetics is sometimes a bit of a gamble and you can't ever erase all chances of a mismarked kit in a litter. But you can certainly increase your chances of mismarked kits by choosing foundation rabbits that do not meet the standard themselves, so always choose carefully.

Breed type is always paramount when choosing foundation stock, but when color and markings are an integral part of the breed's type, don't overlook those characteristics. The future success of your rabbitry may depend upon it. As you gain experience and knowledge, particularly of color genetics, it's possible that you may be able to branch out and utilize rabbits in your program that may not meet the breed's color standard. The key here is to know which colors to mate these rabbits to in order to produce the color you're seeking.

WHERE TO START LOOKING?

Perhaps you will be lucky enough to find a reputable and established rabbit breeder in your local area who can help you to get started with your rabbitry. If you aren't able to find a breeder locally, you will have to do a bit of looking in order to locate your foundation rabbits. Where should you start looking? What are the best places to begin your search?

As you wander around a rabbit show, you might come across a rabbit that simply strikes your fancy. If you're lucky, the rabbit will be available for sale and you'll be able to take it home with you. If you're not lucky, you will have to spend the next month lamenting about the "one that got away."

A-S-K

Let's say that you're looking to purchase a trio of Havanas and have located a very well-respected breeder about two hours from your home. This breeder has produced several rabbits that have been granded and has an excellent reputation for producing quality rabbits. They also happen to have a small selection for sale. You happily make an appointment to visit and hop in the car.

Upon arrival, you're impressed by the clean and airy rabbitry, and the overall quality of the rabbits only solidifies your decision to purchase from this breeder. The owner is pleasant and friendly, answering your questions about Havanas and allowing you to look over all of the sale bunnies. There are five does for sale and four bucks, each of them quite attractive and very healthy-looking. You've been studying the *Standard of Perfection* for Havanas, and all of the terminology is jumbling around in your brain, making your decision very confusing. Which does are the best does, and which buck is the best buck? How can you tell? How do you choose?

Although purchasing decisions are ultimately up to the buyer, it will behoove you at this point to ask for the advice of the breeder. After all, the breeder is the one who has an intimate knowledge of the rabbits and which lines are noted for which qualities. They may be able to recommend one rabbit over another, particularly if you are very precise about your requirements and plans for your rabbits. Additionally, once they've helped you to select two does to purchase, they can also give you recommendations on the buck that would most ideally suit the does in type and pedigree.

The breeder's knowledge and experience can be a definite help when you're attempting to make these important purchasing decisions. Many new breeders appreciate the guidance and assistance that a long-time breeder can provide. One caveat is that a reputable breeder will never try to pawn rabbits off on you. If you feel that a breeder is attempting to persuade you to purchase a sickly or less-than-perfect rabbit against your better judgment, you might want to continue looking elsewhere. Well-respected breeders highly value their reputations and would never do anything to tarnish them, so selling you substandard rabbits for breeding stock would never be in the best interests of their rabbitry. If you feel that a breeder is trying to convince you to start out with rabbits of poor quality, you'd better continue on down that bunny trail to another rabbitry.

Asking a breeder for advice is always a good idea when attempting to choose breeding stock. The knowledge and insight of an experienced breeder can be very helpful for a newcomer who is just getting started with rabbits.

My rabbit show purchase rewarded me with a delightful gift in the form of a litter of kits. An excellent purchase all the way around.

RABBIT SHOWS

If you want to see a lot of rabbits in one place, then a rabbit show is the place to start. In addition to the benefits of personally viewing hundreds (or thousands) of rabbits at once, rabbit shows are also a marvelous place to make contacts. Talk to the exhibitors, ask questions, and find out if they have any rabbits for sale that they have brought along to the show. Even if you don't find exactly what you're looking for at the show, some of the breeders there may have additional rabbits for sale at home who are possibly related to their show winners of the day. Take advantage of the opportunity to meet with as many breeders as you can in order to give yourself the best overview of the rabbits that are available for sale before you make any purchasing decisions. Rabbit shows can be an excellent educational opportunity for any breeder, but particularly for the person who is just starting out with raising rabbits.

NEWSPAPERS

Local newspapers often have rabbits listed for sale in the classifieds section, and these are certainly worth checking into. My personal experience has been that locally advertised rabbits are often mixed-breed rabbits without pedigrees or registration eligibility. They are the kind of rabbits that make excellent pets or 4-H projects, but are not really the type of rabbit that a person would seek for the foundation of a rabbitry. However, don't dismiss the possibility until you've found out for sure. You could be surprised to find that they are purebred, pedigreed rabbits of excellent quality.

WORD OF MOUTH

As the old adage suggests, news travels fast. If you mention to other rabbit enthusiasts that you're looking for a foundation trio of English Angoras, for example, this news can travel by word of mouth and present you with opportunities that you might not have found on your own. It's a small world, especially within the breed specialty clubs. People can help to put you in touch with other rabbitries that just might have the perfect rabbits to meet your needs. In addition, it's also very helpful to receive a personal recommendation from another rabbit breeder who can point you to a rabbitry that is noted for the quality of its stock. When you're just starting out with raising rabbits, you have no way of knowing which breeders have the best reputations for quality, and word-of-mouth contacts can help you to track down a knowledgeable breeder with a good reputation.

BREEDERS

This brings us to our next area of consideration, purchasing from a breeder. It is generally recommended that if you are serious about getting started with quality rabbits, you will want to purchase directly from a reputable breeder. Long-time breeders with good reputations did not reach that status overnight. Years of hard work, enthusiasm, and devotion to the breed; high standards of rabbitry sanitation and record-keeping; and successful marketing are only some of the talents a breeder must demonstrate to become reputable and recommended by other breeders.

BUYING OUT A RABBITRY

Some rabbitries are established and flourish for decades, but others come and go rather quickly. Not everyone who begins raising rabbits is able (or desires) to continue breeding on a long-term basis, due to time constraints, work commitments, health issues, or family obligations. On occasion, entire rabbitries are offered for sale.

This is a case of "being in the right place at the right time." If you're able to locate a breeder who is selling out just as you are looking to get into rabbits, you may have the opportunity to jump-start your breeding program with a ready-made rabbitry. At the very least, you may have the chance to obtain breeding stock that would not have otherwise been offered for sale, and this could include herd sires and senior brood does that are not typically available for sale. Although this isn't an everyday opportunity, it is one worth watching for.

Long-time breeders typically understand the importance of maintaining purebred, pedigreed stock. Chances are that the rabbits that you purchase from such a breeder will be eligible for ARBA registration, which can help to start your new rabbitry out on the right foot. Another benefit of purchasing from an established rabbitry is that their stock is probably already proven on the show table, which gives you better insight into the type and quality of the rabbits produced at the rabbitry. You can also benefit from the experience of the breeder, who has likely been working with the bloodlines in their herd for many years. They can suggest to you the best possibilities for brood does and help you choose a buck that will be complementary to your does, not only in type and physical characteristics, but also in bloodlines. This type of mentoring is extremely valuable as you get started with your rabbitry.

Does this mean that you shouldn't purchase rabbits from John Doe (or would it be John Buck?.) down the street, since he only started raising rabbits last year? Of course not. You should never automatically discount newer breeders and their stock. They may not have years of experience and a breeding program with a long background, but they don't necessarily have rabbits of lesser quality than their long-standing counterparts. Just remember that you will want to be sure of what you're purchasing, and ask lots of questions. Are the rabbits purebred? Do they have pedigrees? Are they registered or eligible for registration?

MAGAZINES

Magazine classifieds are also a good place to shop, although if you're looking at magazines with nationwide distribution, you are more likely to find ads for rabbits from faraway places than from local breeders. This shouldn't necessarily deter you from inquiring, as it's perfectly common to ship rabbits cross-country, and many breeders ship rabbits to distant places on a frequent basis. To start with, check out *Domestic Rabbits* magazine for ads that might interest you, or look for "rabbits for sale" ads in homesteading or small pet magazines.

THE INTERNET

The Internet has introduced a more modern form of rabbit shopping. You can promptly put yourself in contact with rabbit breeders, both near and far, to give you a broader choice of available rabbits. If you type "Dutch rabbit" into a search engine, you will be presented with over five million results, which may give you a glimpse of the vast options that the Internet can provide. You will find links to breeders' websites, links to discussion forums, links to breed clubs, and more information than you could probably ever digest. You will also find a plethora of rabbits for sale, in all of the colors in which Dutch are recognized (and some colors in which they aren't).

As with breeders that you locate via magazine ads, rabbit breeders found on the Internet will probably be some distance from your location. While purchasing long-distance may not be as simple as purchasing from a breeder nearby, you may find that it is well worth any extra hassle if you're able to find exactly what you're looking for. You might be surprised to discover how frequently rabbits are bought and sold across the country, and how easy it is to arrange shipment via airline.

BULLETIN BOARDS

Another place to check when you're shopping for rabbits locally are bulletin boards. As with the local classified ads, it's possible that you may be more likely to find mixed-breed, pet-quality rabbits, but you never know what you might come across. So take a moment to browse the bulletin boards at your local feed store, grocery store, laundromat, or other community gathering places. You could also get creative and place want ads on boards, mentioning that you're on the lookout for "quality Florida White rabbits for a breeding program" and listing your contact information. You might be surprised by a response from a local breeder that you didn't know existed.

If you're looking to buy, you might wind up searching the Internet. E-mails to breeders can be a quick and easy way to find out about rabbits that are available for sale, as well as being an easy way to network.

Let's say that you decided to take a look at some Belgian Hares that you saw advertised for sale. There are many questions that you will want to ask before purchasing any rabbits, and that includes questions about disposition, health, registration, and age.

QUESTIONS TO ASK BEFORE BUYING A RABBIT

1. Can you describe his or her disposition?
2. Has he or she had any health issues, major illnesses, or injuries?
3. Is he or she registered with the ARBA?
4. Has he or she been shown? Any championships or titles?
5. What does he or she weigh?
6. What is his or her age?
7. Why are you selling?
8. Has he sired any litters? If so, how many kits per litter, and how many does were bred?
   Have any of his offspring been successful on the show table?
9. Has she produced any litters? If so, how many kits per litter, and was she a good mother?
   Have any of her offspring been successful on the show table?

CHAPTER 3
HOUSING YOUR RABBITS

Golden light filters through the doorway of this rabbitry on a quiet afternoon. Light and ventilation are essential to the success of any rabbitry, but direct sunlight should always be avoided.

SETTING UP YOUR RABBITRY

This is the part that you may be approaching with delightful anticipation—or with dread. Some people love nothing better than to plan and lay out outbuildings and other farm essentials, while others simply despise the thought of trying to set up and build cages. Whether you are a rabbit equipment enthusiast or would rather sort through pedigrees, this chapter will help to demystify some of the essential pieces of rabbit raising equipment, as well as explain the options you face when initially establishing your rabbitry. Will you use an automatic watering system or water bottles? Will you stack your cages or hang them? Or will you decide that a colony system would work best for your situation? With all of these questions rumbling through your mind, let's begin our discussion of the equipment necessary to set up your rabbitry.

CAGES AND HUTCHES

At the top of your list are cages and hutches, because these will be the foundation of your rabbitry. If you're just starting out with rabbits, you should feel fortunate that you're starting out with such a wide variety of options available to you. Prefabricated cages are relatively inexpensive and easily obtained. Rabbit supply companies offer a vast assortment of cages and hutches in many sizes and structures to help you to create exactly what you need to house your rabbits and suit your own individual needs and situations. Take the time to compare the products offered by various rabbit equipment suppliers and fully examine all of the possibilities before you make a decision. You will be working in your rabbitry every day, so you want the cages to be efficient and easy to work around, as well as safe.

This rabbit breeder is giving a bit of attention to each rabbit as she completes her daily chores. Even though efficiency is important to many people, a few moments of attention to each rabbit is always a good idea.

HOW MANY CAGES?

I suppose the best answer to this is how many rabbits? Generally speaking, each rabbit needs its own cage. Now that's not to say that there aren't instances in which you can house more than one rabbit per cage. In fact, I have two young does sharing a cage at the moment, but only because they are 14 weeks old and best pals. On the whole, bucks need to be housed in their own cages, and does with litters must have their own cages. The only possibility for cage sharing is with junior animals, but you will still want to have these separated by gender to avoid any unexpected early litters. Let's say that you are planning to start with a trio of a senior buck, a bred doe, and a junior doe. In this instance, perhaps you'll decide to start with five cages. This way you will have a cage for the buck, a cage for the bred doe, a cage for the junior doe, and two cages ready and waiting for when the doe's litter is weaned. As your rabbitry grows, so will your number of cages. Perhaps you might decide to plan far in advance and purchase a larger number of cages, enough to house your projected number of rabbits, or maybe you would rather add cages as necessary.

WHAT SIZE?

Cages come in a variety of sizes, and your choice of size will depend on a couple of factors. The size of your rabbits is a major factor. It goes without saying that a Netherland Dwarf will not need as much space as a Flemish Giant. Consider the size of the breeds that you're planning to raise when deciding upon the sizes of the cages you will purchase. As you will notice when reading through any rabbit equipment catalog, you can purchase cages in any manner of size combinations, from the tiniest 18 × 18-inch cages all the way up to 30 × 96-inch cages. In all likelihood, your cages will probably be in the neighborhood of 24 × 24-inch or 24 × 30-inch, unless you are raising one of the larger breeds.

The standard height for cages is 18 inches, but you can also find cages that are 14 inches high, which is perfectly suitable for the dwarf breeds. The 14-inch height allows for the possibility of stacking a fourth layer of cages to maximize the space of your rabbitry.

Aside from the size of your rabbits, you might also consider using larger cages for does with growing litters and smaller cages for junior bucks

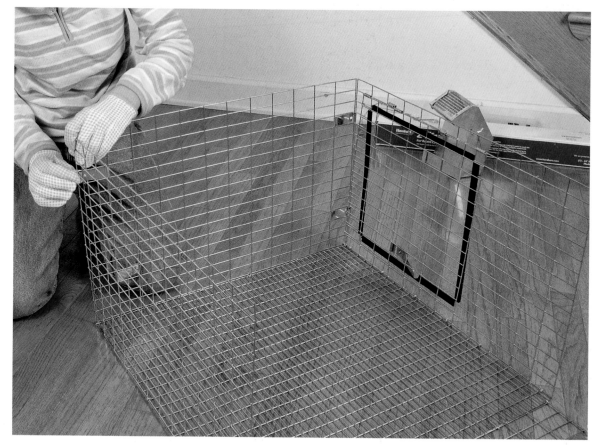

An all-wire rabbit cage can be purchased as a kit and comes partially assembled with all of the parts pre-measured and ready to go. If you are slightly more ambitious, you can create your own wire cage from scratch and save money but add time to the process. One of the attributes of the prefabricated wire cages is their ease of setup and the fact that they can be assembled in a very short period of time.

and does that are living alone but not yet fully grown. Experiment a bit within your rabbitry until you find out what works best for you.

CAGE AND HUTCH VARIETIES

Generally speaking, "cage" is the term used when referring to all-wire rabbit domains, and the word "hutch" is used to describe an enclosure that is partially or entirely made of wood. Wooden hutches were much more commonly seen in years past. In fact, they were the standard type of rabbit housing for many years. However, today they are rarely seen, as the vast majority of rabbit breeders have switched to all-wire cages for a number of reasons. The ease of cleaning and disinfecting is a major reason that all-wire cages are now the standard; also, rabbits cannot chew on wire cages as they can on wooden hutches. In addition, wood has the unfortunate tendency to rot, particularly with years of use and

continual exposure to rabbit urine. Today's all-wire cages are easy to assemble, incredibly easy to clean, easy to work around, and generally considered to be the one-and-only option for rabbit owners. They are also relatively inexpensive and last for many years. Hutches, of course, have the advantage of aesthetics. A wooden hutch with a bunny inside is always a pleasing sight for the first week or so. Once the rabbit has sprayed the walls with urine and chewed on them, it doesn't have quite the same aesthetic appeal. All-wire cages, as long as they are kept clean, always look nice.

Rabbit equipment manufacturers have developed different types of all-wire cages, with one of the main options being doors that swing in and up or doors that swing out. Having worked with both types, I have to say that in my opinion, the doors that swing out are much easier to work around. An advantage to the doors that swing in and up is that

Many all-wire cages have doors that swing inward and up toward the top of the cage. The advantage to this system is that the rabbit cannot push the door open from the inside. However, many people prefer their cage doors to swing out.

the rabbit cannot push its way out of the cage since the overlapping part of the door is on the inside. However, I find them hard to work around, because if a rabbit is sitting on the inside of the door and you need to open it, you are opening the door into the rabbit, which isn't always interested in moving. It can be a bit frustrating to try to move a nine-pound Rex that has positioned herself in front of her cage door when you can't get the door open to move her. For these reasons, I prefer swing-out doors.

Another choice to be made is between hanging cages and stacking cages. There are several advantages to a hanging cage system. First of all, the fact that the cages are off of the ground helps keep your rabbits safe from raccoons, rats, or other troublesome pests. Secondly, the fact that the cages are suspended in the air allows you to decide whether or not to utilize trays. If you choose not to use trays underneath the cages, then all of the droppings will fall through directly to the ground. Many rabbit breeders prefer this setup for a couple of reasons. It eliminates the need to regularly clean trays, and it also allows for making easy compost right underneath the cages. (See more on this in Chapter 6.)

Removable plastic trays are incredibly easy to clean. A warm water rinse easily washes away any residual manure or urine, as well as any hay or pellets that have been spilled through the cage floor.

Stacking cages are also advantageous, for different reasons. With stacked cages, you can maximize space, which is a definite plus if your rabbit shed or building is on the small side. If you stack your cages in sets of three, you can quickly triple the number that you could keep in the same space if you used the hanging cage method. This efficiency allows you to expand your rabbit herd to a level that you could not accommodate if you had room for only one layer of cages.

It's recommended that you do not place your cages on the floor, as it would be too easy for predators or rodents to access your rabbits. If you don't want to hang your cages, you may want to build a base for them to sit on so that they are a few feet off of the ground.

A final decision that you will need to make is whether to purchase metal trays or plastic trays for your cages. For years I had my rabbits in cages with metal trays, but I've recently switched to the newer plastic trays. I've been amazed at the difference that such a slight change can make. The plastic trays are the winner, hands down, in my estimation. Not only are they considerably easier to clean and keep clean, but they are lightweight and easy to carry, and they don't have sharp edges or rust. Given the choice, I will always choose the plastic trays over the metal ones.

Wooden hutches can be extremely attractive, but they usually aren't a practical choice for a rabbitry. The wood portions can be easily damaged by urine spray and by chewing. In addition, the fragile legs of some hutches can be broken if a predator attacks.

These homemade cages utilize a popular design with a slanted drop pan between the layers of cages. This replaces the need for a plastic or aluminum tray underneath each cage because the droppings and urine slide down the slanted drop pan and onto the ground. This makes for easy cleanup and eliminates the need for washing trays.

These are all-wire cages that have been reinforced with wooden frames. This design makes the cages more substantial and sturdy, and the fact that they are raised off of the ground gives additional protection against predators such as weasels, raccoons, and snakes. Stacking cages, such as the ones on the left, can really maximize your space and allow you to double or triple the number of cages that can be kept in the same area.

Your cages can be kept in a row, which allows your rabbits to see one another and benefit from the visual companionship. Any rabbit that shows the slightest indication of illness should be promptly removed from the main group of cages and housed separately. This way, the germs cannot be transmitted between rabbits through cage walls.

Smaller breeds require less cage space than the larger breeds, so if you are working with a small area for your rabbitry, you might want to consider working with one of the smaller or dwarf breeds rather than choosing a larger or Giant breed.

CAGE ACCESSORIES

We'll discuss additional equipment (feeders and waterers) in a moment, but for now, let's go over some of the other accessories that you might consider when setting up your cages and getting started with your rabbits.

Urine guards are used by some breeders, as they can help to keep your rabbits from spraying the walls and floor outside of their cage. You can choose between plastic or metal urine guards, but I've found that the plastic is much easier to keep clean.

Floor mats (also known as resting boards) can be very much appreciated by your rabbits, as they allow the rabbits to sit on a smooth plastic liner that sits between their feet and the wire cage floor. Because the floor mats are slotted, the droppings slide right through, keeping the resting area tidy. However, if a rabbit does happen to dirty the floor mat, it is plastic and easy to wash.

Door liners can be placed on your cage doors to eliminate the possibility of scratching yourself or your rabbit on the sharp edges of the door. These are inexpensive and handy devices that can save you from an uncomfortable scratch.

A small barn can be an excellent building in which to house your rabbit cages. It will give protection from adverse weather conditions and also protection from most predators.

This litter is being housed in a large wire cage. The advantage of this particular cage is that its door opens from the ceiling, which makes it very easy to reach in and access your rabbits.

INDOORS? OUTDOORS? OR A LITTLE OF BOTH?

It's one thing to purchase cages, but it's another to decide where to keep them. Rabbit breeders house their rabbitries in a wide variety of shelters, and some aren't in shelters at all. It would be hard to say that one method is better than another, since so much depends on the individual circumstances and the size of your rabbitry.

A small barn or garage is often converted into a rabbitry. One benefit of these dwellings is that they provide shelter from all types of adverse weather conditions, from wind and rain to heat and humidity. In addition, an enclosed building can go a long way toward protecting your rabbits from wild animals or dogs that might try to attack them. Of course, an enclosed building offers its own challenges as well, since you will have to consider lighting, ventilation, and a water source, along with the possibility of heat to keep the building warm during the winter months.

This wooden hutch comes from a kit that only requires the use of a few common tools. Wire cages can be more quickly assembled, but they perhaps lack the charm of a wooden hutch.

Young rabbits that are newly weaned can often be housed in the same cage together. However, after approximately 8 to 12 weeks, you will want to separate the bucks from the does. Additionally, the bucks may not be able to share cages due to potential fighting.

Another option is to build an indoor/outdoor type of shelter with a covered roof and open sides. The perimeter of this type of structure is sometimes enclosed with wire (similar in appearance to a chicken coop), and the cages are either hung from the ceiling or built onto platforms that raise them off of the ground.

It's possible to have your cages outdoors, but typical all-wire cages are generally not sturdy enough to withstand the threat of predators. Outdoor cages are usually converted into hutches and reinforced with wood so that they are stronger and harder to tip over and break. You must consider the weather elements when placing cages outdoors, as they should not be in direct sunlight or exposed to inclement weather.

WATER

Water is such a seemingly simple topic, but one of such monumental importance. Water is one of the key components to any successful rabbitry, and it's not always easy to decide which type of watering system will work best for you and your rabbits. You want a system that is reliable and easy to manage, as well as clean and efficient. Let's take a quick look at the various types of systems for watering rabbits.

WATER BOTTLES

One of the most popular options, and my personal favorite, are water bottles. These are plastic bottles with a nozzle that screws onto the bottle opening and hangs on the outside of the hutch on a metal wire. The nozzle reaches inside the cage so that the rabbit can access it. The metal ball at the end of the nozzle prevents the water from leaking out, but when the rabbit licks the nozzle, the ball moves back and the rabbit is able to drink. Water bottles have many advantages, including the fact that you can carefully monitor how much your rabbit is drinking (or not drinking, as the case may be), and the fact that your rabbit cannot spill the water. The only disadvantage to the bottles is that they can occasionally leak if you haven't set the vacuum seal on the nozzle, but this is an unusual occurrence and they typically work very well.

Water bottles are a very common fixture of rabbitries. The water is released when the rabbit licks the nozzle, which dislodges a ball and allows the water to flow. When the rabbit stops licking the bottle, the ball falls back into place and blocks the water flow.

WATER CROCKS

Perhaps a more traditional type of water source than the water bottle, the water crock is used by many rabbit breeders. Its simplicity is one of its main attributes, as it is quite simple to wash and refill a small crock each morning as opposed to taking down a water bottle, unscrewing the nozzle, and washing, rinsing, and refilling it each day. The weight of the water crock typically prevents spillage, but I've had rabbits that were positively determined to tip over their heavy crock and always succeeded in doing so if they truly desired. This is the drawback of the water crock system, along with the fact that because the crock is on the floor of the cage, the rabbit will sometimes use it for unsanitary purposes. It adds a good amount of chore time each day if you have to wash out urine or droppings from the water crock.

Another option, and one I've had better success with than with water crocks, are the small plastic clip-on cups that are sold through rabbit equipment suppliers. These small cups fasten onto the wire of the cage wall and are often used in rabbit carriers at rabbit shows. However, I like to use them at home, particularly with rabbits that aren't fond of using the water bottle. They are lightweight and easy to wash, but because they snap onto the cage wall, it is very difficult (nearly impossible) for a rabbit to pull one off of the wire. The height and depth of the cups also discourage the rabbit from using them as a litter box. The disadvantage of the cups is they aren't huge and don't hold a lot of water, which means that you will be refilling them often, particularly with larger breeds.

AUTOMATIC WATERING SYSTEMS

For larger rabbitries, an automatic watering system can be an incredible time saver. An automatic watering system is designed to provide water continually so that your rabbits never run out. These systems are a more expensive option than water bottles or crocks, but many breeders feel that the expense is well justified in exchange for the time saved in refilling such containers. The drawback to this type of system is the fact that you cannot monitor the water intake of individual rabbits, and if the system were to malfunction and stop water production, you might not notice immediately, which would leave your rabbits without a water source. However, many commercial rabbitries have nothing but praise for their automatic watering systems.

Crocks come in a variety of sizes and types, but you'll want to use heavy crocks that are harder for rabbits to spill. The lightweight stainless steel crock on the top is probably a poor choice for a rabbit. A better choice would be one of the heavier crocks on the bottom.

OTHER WATER OPTIONS

I've settled upon my own ideal setup, which works for me only because I keep my rabbit herd at a level that's easy to manage. If I had 20 does and 8 bucks, I'm sure that I wouldn't be able to handle providing my water this way, but for now it works perfectly. Each cage has a water bottle, and the majority of the cages also have a small water cup as a backup plan. I like my rabbits to have more than one source of water, so that they have an option. Some prefer the water bottles, some prefer the small cups. I haven't been able to detect any rhyme or reason as to which rabbit chooses which water source, but I have noticed that some rabbits are far more likely to drink from the cup than they are to drink from the water bottle. Therefore, I like to keep both options in front of them. That way I know that if they happen to drink the cup dry, they still have an entire water bottle to drink from if necessary. Or if the nozzle on the water bottle were to malfunction (it's never happened, but you never know), then the rabbit would still have a water source.

I firmly believe that water is one of the key elements to any rabbitry's success, and that's why I like to go the extra mile toward keeping fresh, clean water in front of my rabbits 24 hours a day, seven days a week. As we will cover in Chapter 9, water is even more important for nursing does. They need ample amounts of fresh, clean water while they are nursing their litters.

This is one of the most common types of rabbit feeder. The feeder hangs on the outside of the cage with only the feeder portion inside the cage. The pellets are added through the flip-top lid.

FEEDERS

You will need a feeder for each hutch in which to place your rabbit's pellets. Many breeders also use their feeders to dish out other delicacies, such as dandelion greens or bread. What kind of feeder is the best choice?

Generally speaking, you have a choice between an outside feeder, an aluminum feed pan, or a crock. As with the water crock, your feed crock will need to be heavy enough that your rabbit cannot easily tip it over and deep enough that your rabbit cannot scratch out the feed. In either case, unnecessarily spilled feed is an unfortunate waste.

Outside feeders eliminate any chance of scratching or spilling and are designed so that your rabbit cannot leave urine or droppings in the feeder. To use an outside feeder, you will have to cut the wire on the hutch to allow room for the base of the feeder to fit inside the hutch and leave the remaining portion of the feeder outside the hutch, where you will add each day's feed without having to actually enter the hutch yourself. This is a very common type of feeder. You can purchase outside feeders with either solid bottoms or wire screen bottoms; each type has its merits. The solid bottomed feeder is ideal if you're feeding a type of pellet that is very tiny or if you're supplementing with seeds or other small foods. The feeder with the wire screen bottom is also very popular because tiny bits of feed dust (known as "fines") can slip through, and you will never have to take the time to clean them out.

One disadvantage to outside feeders is that they can be somewhat sharp on the edges and corners, and the potential is there for a rabbit to poke or injure itself while eating. These instances aren't common, but they do occur, so you should always be aware of sharp areas that could potentially harm one of your rabbits.

I've had good success with small aluminum feed pans. They clip onto the hutch wall, and you can easily raise or lower the feeder depending on each rabbit's size. They are easy to clean and fill, and it's very easy to empty out any fines or leftover uneaten feed. The only drawback to this type of feeder is that it is possible for the rabbit to use it as a litter box. However, this problem is easily remedied by moving the feeder to a different part of the cage or by lifting it an inch or two off of the cage floor so that it is low enough to eat from, yet high enough to be out of reach of any urine or droppings. The rounded edges of aluminum feed pans make them a very safe choice.

Another option is the snap-on plastic cups that are often used for water. These can work double-duty as feeders, especially in carriers if you're taking your rabbits to a show. The drawback is that these cups hold very little feed, but this isn't an issue if you're feeding one of the smaller or dwarf breeds, since they usually aren't consuming vast amounts of pellets.

HAY RACKS

Kudos to the person who invented the hay rack, as it is one of the most useful pieces of equipment that you can add to your rabbitry. Instead of placing your rabbit's daily hay ration on the floor of the hutch, where the rabbit will inevitably trample and soil the hay, you simply hang a hay rack on the outside of the hutch wall and fill it with hay. This setup allows your rabbit to pull the hay through the wire and prevents him from stepping all over it or dirtying it with urine or droppings. In this way, you will prevent a great deal of waste and still allow your rabbit access to plenty of hay. As with pelleted feed, you won't want to provide more hay than your rabbit will eat in a day's time.

A hay rack is an excellent way to provide hay. Not only does it reduce waste, but it also reduces your rabbit's risk of a condition known as "urine burn." This problem is also known as "hutch burn," and can be caused when a rabbit sits on soiled hay for a long period of time. Hay racks prevent the buildup of soiled hay on the cage floor.

A nest box is typically viewed as an essential item of equipment for anyone with breeding does, but some rabbitries provide nest boxes for each and every one of their rabbits, including bucks and juniors. Rabbits enjoy having a dark place to hide in and something to sit on, and a nest box provides both.

Urine guards can be placed around the bottom edges of your wire cages, which helps keep the rabbits from spraying urine outside of the cage.

BEDDING

Most rabbit breeders do not add bedding to their rabbits' hutches. In order for the design of the wire floors to work properly and allow the urine and droppings to fall through to the pan below, the floor cannot be covered by any type of bedding. There are a couple of options to make the wire floor a bit more comfortable on your rabbits' feet. A small wooden board can be placed in one corner of the hutch, or you can purchase plastic floor mats for the wire floor. These floor mats have gaps that allow the urine and droppings to fall through, yet they help keep the floor more comfortable for your rabbit at the same time. These are particularly helpful for the giant breeds, which have an increased tendency to develop sore hock (a condition that causes ulcers on the rabbit's feet).

Bedding can be added to the tray underneath the cage. A small amount of wood shavings can help soak up urine and may also help reduce hutch odors. Nest boxes must have bedding added. Many people line the bottom of the nest box with wood shavings and cover the top with hay or straw. Others use shredded newspaper covered with hay or straw. Hay can sometimes be a bit softer than straw, which makes it a more comfortable choice for newborn kits, but this can vary somewhat depending upon the types of hay and straw available.

CARRIERS

They are not an object that you will use every day, but a couple of rabbit carriers are very useful items to have around your rabbitry and are vital pieces of equipment to invest in. If you're planning to show your rabbits, carriers are a must-have. They're easy to transport, they keep your bunnies safely enclosed in a small space, they have handy pans underneath to catch droppings, and they allow you to easily have your rabbits in the right place at the right time on show day. You can purchase individual carriers, double carriers, three-in-one carriers, and so on all the way up to six-unit carriers, depending on your needs.

Carriers are a wise investment even if you're not showing. If you're delivering a rabbit to its new home, picking up a new rabbit you've purchased, or taking one to a veterinarian for some reason, a carrier will make your trip much easier. Carriers are also handy to have if you want to thoroughly clean your rabbit's cage, as the carrier will give you a safe place to place the rabbit while you work. Carriers will last for years and their prices are actually quite reasonable when you consider this fact.

DAILY CARE OF YOUR RABBITRY

The cages are set up, the waterers and feeders are filled and working well, the bunnies are happily enjoying themselves in their new abode, and you're settling back to relax. Or are you? Your daily chores in your rabbitry are a bit more complicated than simply refilling feeders and waterers, as you will soon find out if you haven't already. In order to maintain a properly managed rabbitry, you will need to pay close attention to many other areas in order to keep things in smooth running order.

Whether or not you decide to pursue the idea of showing your rabbits, you will still want to have a rabbit carrier or carriers to transport them as needed. The carrier pictured here has space to carry four rabbits, side-by-side.

Always double-check your latches. A partially closed latch can spring open and allow your rabbit the opportunity to escape. Always avoid this predicament by carefully checking to see that each cage is closed when you're finished working.

The pull-out tray seen on this hutch is narrower than the ones that are typically found on wire cages. Frequent cleaning will be required to reduce odors or buildup of droppings.

CLEANLINESS

Cleanliness is paramount to the success of any rabbitry, so pay scrupulous attention to this important subject. Frequently wash your water bottles, water bowls, and feeders. Daily rinsing will usually suffice for your waterers, but they should also be thoroughly cleaned at least once a week. Wire cages stay relatively clean, but you will still want to take the time to disinfect the cage on occasion. Take the time on a weekly basis to remove any loose bits of hay or any droppings that are stuck to the cage wire. Keep a close eye on your nest boxes, and don't assume that your does will be using them only for nesting. Many does decide that

a nest box makes a handy litter box. You certainly don't want any kits born into a dirty nest box, so check it frequently in order to remove any manure or soiled bedding.

Your hutch trays must be cleaned at least once a week, if not more often. If you leave these pans for too long, the ammonia smell may become strong. You don't want any major buildup of droppings, either. Pull out the pans and carry them over to whatever location you have designated for your rabbit manure. Take a small board (a 1 × 4-inch board works very well) and scrape each pan clean. Follow this up with a warm water rinse, and you'll be all set. Some breeders like to sprinkle a bit of baking

61

This young rabbit is having difficulty standing on the wire of the cage. Her hind feet are slipping through the wire. A resting board or floor mat might be a helpful addition to the rabbit's hutch and give her a comfortable place to rest, while still allowing for the droppings and urine to fall through to the pan below.

VANODINE

Vanodine is one of the most popular disinfectants with rabbit owners and is prized for its qualities as a disinfectant and for its safety. Unlike many other cleaning products, Vanodine is reported to be perfectly safe for use around animals and will not harm them even if it is consumed. Vinegar is another product that is commonly used for cleaning rabbitries. It is certainly an all-natural product that appeals to many rabbit owners. Other rabbit breeders prefer using a product such as Lysol for disinfecting their cages and nest boxes.

soda in the bottom of the pan to reduce odors, and others prefer to line the pans with newspaper for easier cleanup. I've tried adding newspapers and found them to create more mess than they prevent, so I've foregone that idea at present. Some people also line the tray with wood shavings to help absorb urine and reduce ammonia odors.

VENTILATION AND LIGHTING

Ventilation and lighting are two key areas to consider when setting up your rabbitry. Either of these topics might be easily overlooked, but they are truly important aspects. Rabbits need fresh air, so air movement within the rabbitry is crucial, although access to drafts should definitely be discouraged. There is a difference between receiving fresh air and being continually in the path of relentless wind. If you don't plan for proper ventilation and allow for fresh air to continually circulate within your rabbitry building, you could damage your rabbits' respiratory systems and put them at a higher risk of contracting pastuerella ("snuffles"). Carefully consider the structure and layout of your rabbitry and determine how you will best allow for the movement of fresh air without the problem of frigid drafts. This may involve opening windows, running fans, or whatever else is necessary to make sure that the stale air in the rabbitry is replaced by clean, fresh air.

Lighting is vital for two reasons. The first is that it's frustrating to try to work efficiently in a building that is poorly lit. Therefore, ample lighting throughout your rabbitry will completely illuminate your daily chores and make them much easier and more pleasant. Secondly, adequate lighting can help increase the litter production in your rabbitry during the winter months. If you can emulate the longer days of summer through artificial lighting, your rabbits will more than likely produce more consistently during the winter months than they would otherwise. By keeping your rabbitry well ventilated and well lit on a daily basis, you are helping to decrease your rabbits' chances of illness and increase their chances of pregnancy. Sounds like a good trade-off, eh?

This rabbit is listening intently to a noise from above. Rabbits are very easily frightened by noises, particularly squeaking noises or loud squeals.

A DAILY DOUBLE-CHECK

At the end of the day when I'm finished with the evening feeding, I like to make a quick double-check of everything, so I walk back through my rabbit building and make sure that I haven't inadvertently forgotten something. It takes only a moment to make sure that all of the water bottles are filled, all of the feeders have pellets, and all of the cage doors are latched. It's an excellent feeling to know that all is well with the rabbits and everyone is happily enjoying their dinner.

SPECIAL CARE IN WINTER

As with any type of livestock, winter weather conditions can present certain challenges in providing proper care. Rabbits are fairly adaptable to cold temperatures—much more so than they are to hot temperatures—but special care needs to be taken in the wintertime as well.

Fertility is believed to be somewhat decreased during the winter months. This is a natural occurrence, as winter is certainly not the optimal time for rabbits to be born in the wild. But in the controlled environment of your rabbitry, it's very possible that you will want to try to raise litters all year round. In this case, you may need to take special steps in order to continue producing litters at the same rate as in the warmer months. Light is extremely important. The shortened winter days and the reduced hours of daylight play a part in the decrease in fertility, so you can counterbalance this by providing light in your rabbitry for at least sixteen hours per day. This simulates the amount of daylight during the summer months.

Water will be one of your main concerns during the winter months because unless your rabbits are housed in a heated building, it's very likely that you will be faced with the problem of frozen water. This is true whether you use water bottles or water crocks, and you may also be faced with the problem of the bottles or crocks cracking as the ice expands. You may find that it's easier to switch to water crocks during the winter so that you can swap out the frozen ones with fresh ones and bring the frozen ones indoors to thaw. It's very important that you provide fresh water at least twice a day during

In hot weather, it's a good idea to freeze small bottles of water. You can provide these frozen bottles to your rabbits in their cages. These bottles give your rabbits a cool place to rest near, which is particularly important because rabbits often suffer in hot, humid weather.

Record keeping is important regardless of whether you house your rabbits in cages or in a colony setting. This doe is easily identified by the tattoo in her ear. Some breeders decide to forego record keeping when they establish a colony and figure that pedigrees are unimportant when their aim is to raise meat rabbits. You should never underestimate the importance of proper record keeping because this will reap dividends for you later on.

the winter months, preferably more often. Some breeders like to provide warm water for their rabbits during the winter, as it is a bit more palatable than ice cold water, but others feel that warm water has the potential to cause digestive upset.

It's always important to provide protection for your rabbits from the harsh winter elements so that they can get out of the bitter winds and blowing snow. If you don't have an indoor building for your hutches, you can improvise by placing them in a secluded area that is out of the wind and securing a tarp along two or three sides of each one. All the same, it's important to maintain good ventilation for your rabbits, even in the winter. Fresh air is a must.

You should never overlook the fact that it takes a lot of energy for an animal to stay warm in the wintertime. They are burning many more calories during frigid temperatures trying to generate enough body heat. Therefore, your rabbits' hay ration should definitely increase during the winter, as they will need the additional calories to maintain their weight.

SPECIAL CARE IN SUMMER

As with winter, summer weather presents its own unique set of challenging circumstances. In the summer months, your most pressing concern will be protecting your rabbits from the heat. Rabbits do not sweat, and therefore they can easily suffer during the warm summer months if you do not take steps to make them as comfortable as you can.

Shade is obviously very important so that your rabbits are able to escape from the direct sunlight. Always make sure that your rabbits have somewhere to sit that is out of the heat of the sun.

Ventilation, whether in the form of open windows and doorways or electric fans, is another area to consider. Air movement and fresh air are necessary at all times of the year, but particularly so during the hot summer months. High humidity levels are very stressful to rabbits, so strive to maintain a well-ventilated area with low humidity.

If it is particularly hot and humid and your rabbits seem to be uncomfortable, it's a good idea to freeze a few water containers (recycled one-liter plastic soda bottles work really well) and place them in each cage. This way, the rabbits can lay next to the bottle and enjoy the cool, refreshing feeling of the ice next to them.

It is a noted phenomenon that bucks may experience temporary sterility after a lengthy period of extremely hot weather. Because this temporary sterility can last for up to three months, you will want to take special care to prevent the condition in the first place. If you can keep your bucks in a particularly cool location, you may want to consider doing so, especially if you want to continue producing litters during the late summer and early fall.

RAISING RABBITS IN A COLONY

For years, rabbits have been housed in the same manner: in hutches with wire walls and floors, with pans underneath to catch the urine and droppings. The hutches are lined up in long rows under the rabbitry roof, each one accompanied by its own individual water bottle and feeder. This has become the standard way to raise rabbits and is a sort of "tried-and-true" method, if you will.

Quite recently, rabbit breeders have been rethinking the way that rabbits are raised. There's no dispute that housing rabbits in hutches is a perfectly acceptable method with wonderful results, but some breeders have begun to experiment with trying to

Snakes are relatively harmless to your cage-housed rabbits, but they are a troublesome predator for rabbits kept in colonies. The snakes can easily sneak into the colony and attack your kits.

The opportunity for rabbits to obtain fresh green foods is another reason that many breeders prefer the colony style of rabbit keeping. The incidence of enteritis can be somewhat higher in colony-raised rabbits, so you must carefully consider this drawback in comparison to the benefits.

Rabbits that are raised in colonies have ample time for exercising. Even if you are raising rabbits in wire cages, you can still set up a small fenced area outdoors and allow your rabbits the occasional opportunity to exercise and play.

house their rabbits in a different way. Following closely on the heels of the "free-range chicken" and "grass-fed beef" movements, these rabbit breeders are moving toward the idea of "colony raising" their rabbits.

Colony raising is gaining popularity because some breeders feel that it is a more natural approach than the traditional hutch method. Colony-raised rabbits have the ability to move about without the restrictions of being confined to a hutch. In many cases they are able to dig and burrow and nibble on fresh greens, which are very natural pursuits for a rabbit.

Because the practice of raising domestic rabbits in colonies is still in its infancy, there are a wide variety of opinions on how to go about it. Some of these differences of opinion pertain to the resources that each individual breeder has available, some pertain to the number of rabbits being placed in the colony, and some are simply a matter of each person's perceptions of how the idea can best be carried out.

Outdoor colonies are the most natural, but many people have experimented with setting up indoor colonies. This is often accomplished by using an empty 10 × 10-foot or 12 × 12-foot stall, although these dimensions cannot support very many rabbits. Using a stall is, however, a good way to get your feet wet with colony raising to see if it suits your purpose.

There are many options for setting up an outdoor colony, but the first area of consideration is putting up enclosure fencing. This is a vital step in order to keep your rabbits in and the predators out. Chicken wire is generally considered to be unacceptable, as the holes are too large and baby rabbits can potentially escape. Hardware cloth is sometimes used, as the holes are small enough that a rabbit is unable to squeeze through it. The fencing is usually extended at least 24 inches underground in order to prevent the rabbits from burrowing underneath the fence. Alternately, the entire bottom of the colony area can be covered with wire mesh or hardware cloth and then covered with a thick layer of dirt, which will further prevent the accidental loss of a rabbit through digging out.

One of the reasons that some breeders prefer the colony setting is because it allows their rabbits the opportunity to burrow. Digging is a natural instinct for rabbits and is something they are unable to do in a wire cage. This rabbit is enjoying herself as she digs in this colony.

Some breeders like to run a strand of hot wire along the perimeter of the fence, 3 or 4 feet off of the ground, to discourage any predators from attempting to enter the colony. Additionally, some people cover the entire top of the colony with screen or wire to protect the rabbits from being attacked from above by hawks or other predator birds. Breeders occasionally set up colonies that display a more building-like appearance, along the lines of a chicken coop, with wood framing and a complete enclosure of wire on all sides.

There are differing opinions on some aspects of raising rabbits in a colony. Some prefer to keep the buck separate from the bulk of the colony and allow the does to live and raise their kits together. Others prefer to keep the buck with the rest of the colony. Some people remove the does from the colony when it is time for them to kindle to let them raise their litter in the privacy of a hutch.

There are several advantages to colony raising, which is why increasing numbers of rabbit raisers are switching to this type of housing. The natural aspects are of course a major attribute, but so is the time saved on cage cleaning. Although colonies need to be cleaned, doing so is usually not as labor intensive as in a hutch system. Rabbits can exercise to their heart's content when raised in a colony, and they can enjoy the companionship of other rabbits rather than spending their days alone in a hutch. Additionally, you can keep more rabbits in a colony setting than you might be able to in a hutch setting, giving you wider options to pursue different courses in your breeding program or to raise more rabbits for meat purposes if that is your goal.

Of course, there are disadvantages to raising rabbits in colonies. It is impossible to completely eliminate the danger of predators, from raccoons and weasels to wolves. Snakes and rats are particularly troublesome pests, and their small size makes it easier for them to squeeze their way into the colony. Some rabbit breeders who have attempted colony raising have reported a substantial loss of kits due to these dangerous predators.

It's also possible that your rabbits may find a means of escaping from the colony, no matter how carefully you may try to keep them safely contained. You may find that it is more difficult to handle your rabbits once they have been raised in a colony setting, and you may find that it is considerably more difficult to catch them in a colony than it is to catch them in a hutch.

Sanitation is another area of concern. An all-wire hutch can be easily cleaned and disinfected, but it can be trickier to try to sanitize an outdoor area in the unfortunate event that a communicable disease were to crop up in your colony.

It's also necessary to consider the importance of protection from the elements. Colony-raised rabbits must have access to some type of shelter in order to be protected from wind, rain, hail, and snow, as well as the intense heat of the afternoon sun. This can be as simple as a small three-sided covered box or a small plastic hut. Experiment to see what works best for your setup and situation. Similarly, you will want to provide your colony with some type of covered feeder to protect the feed from being damaged by rain or snow.

Raising rabbits in traditional hutches is still by far the most popular method of rabbit housing, but the idea of colony raising is certainly gaining a lot of attention as an alternative method.

One of the problems with colony raising is that you may have difficulty in catching your rabbits when necessary. Rabbits enjoy their freedom and are very quick at scurrying away. Handling your rabbits frequently will help to keep them used to being caught and less fearful of you.

Sometimes it's not wild animals that cause trouble in your rabbitry. Your own pets can occasionally cause destruction, as well. In this case, a Chihuahua has stationed himself outside of this colony of rabbits. Always keep a close watch on your pets to make sure that they don't cause harm to your rabbits.

CHAPTER 4
FEEDING YOUR RABBITS

Wouldn't it be delightfully simple if you could just toss a few carrots in each cage and head off to work for the day? No hay, no pellets, no measuring. Feeding rabbits isn't quite as easy as that, but it really isn't terribly difficult either. Once you've established a feeding program that works for you and your rabbits, your daily feedings should be simple and straightforward.

This Holland Lop is munching on some alfalfa hay. My personal experience is that my rabbits vastly prefer alfalfa hay to grass hay, but some breeders have reported incidences of diarrhea or other digestive problems in their rabbits after feeding alfalfa.

Although you can't tell by looking, these three varieties of pellets have varying amounts of protein. The pellets on the left are a basic 16 percent protein rabbit feed, which is suitable for most rabbits. The pellets in the middle contain 17 percent protein and have less fiber than the previous brand. The pellets on the right are an 18 percent protein feed and are specially formulated for promoting show condition.

BASIC NUTRITION

As with all types of livestock animals, rabbits need a well-rounded diet that fulfills all of the necessary requirements for basic nutrition. This means that their diet must include ample sources of fiber, protein, vitamins, and minerals. With the wide variety of pelleted feeds that are available today, it's easier than ever to provide proper basic nutrition for your rabbits, but a well-rounded diet also contains other important elements: water and hay. Let's discuss the components of a healthy diet for your rabbits.

PELLETS

If you ask a group of rabbit breeders what they feed their rabbits, I can pretty much guarantee that the vast majority of them will reply that they feed some type of pellets. Pelleted feed has become the industry standard for many reasons.

Using pellets takes the guesswork out of providing a diet for your rabbits. With a quick examination of the tag, you will know exactly what is included in each and every feeding. Pellets are easy to measure, which ensures that you are able to feed the same amount each day, and they are an easy way to provide all of the nutrients that your rabbits need for optimum health and growth.

There are many companies that produce rabbit pellets, and it seems that each type is supported by legions of rabbit breeders who believe strongly in the benefits and quality of their chosen product. It can be a little overwhelming to flip through the latest *Domestic Rabbits* magazine and read through the ads for rabbit feeds. Each one is endorsed by rabbit breeders, exhibitors, and judges, and each feed is seemingly better than the last. How, then, do you determine which feed will best suit your needs?

Your location is undoubtedly one consideration. Not all feed companies have distributors all over the country, so your choices may be limited somewhat by your region and what is available locally. With that in mind, it might be a good idea to talk with some of the successful breeders in your area. Ask them about their choice of feed and whether or not they are satisfied with it. I recently discussed feed with four breeders and received three different answers with regard to their suggestion for "the best feed." In fact, one person highly recommended the very feed that another person specifically said not to use. How's that for contradictory advice?

Even after you have settled upon a brand of feed to purchase, there is still the decision of which type of feed will best suit your needs. Each manufacturer typically produces more than one type of feed with subtle differences in protein and fiber amounts. For instance, protein amounts in pelleted feed can vary from approximately 12 percent all the way up to 18 percent. Many breeders settle upon a feed in the middle, with 16 percent protein. Eighteen percent protein is considered by some breeders to be unnecessary at best and unhealthy at worst. Some say that this type of feed can cause diarrhea. Others feel that a feed with 18 percent protein gives an added boost to young rabbits that are growing rapidly. Rabbits that have reached maturity and are not bringing up a litter (bucks and non-lactating does) may be just fine on a pellet with a lower protein percentage of perhaps 14 percent.

Daily Feeding Amount

Rabbit (Small Breed).................................. ...1/4 to 1/2 cup
Rabbit (Large Breed)................................. ....1⁄ to 1 cup

GUARANTEED ANALYSIS

Crude Protein (min.)... 17.0%   Calcium (min.) 0.6%
Crude Fat (min.)............ 3.0%   Calcium (max.) 1.1%
Crude Fiber (min.)...... 14.0%   Phosphorus (min.).......0.4%
Crude Fiber (max.)...... 18.0%   Salt (min.)................ 0.25%
Moisture (max.).......... 12.0%   Salt (max.)................0.75%
                                                  Vitamin A (min.)..5000 IU/lb

INGREDIENTS

Dehydrated Alfalfa Meal, Ground Oats, Wheat Middlings, Dehulled
Soybean Meal, Ground Wheat, Ground Corn, Dried Cane Molasses,
Dicalcium Phosphate, Corn Oil, Calcium G...

Always check the label on your feed bag. As you can see, this type of feed contains 17 percent protein and between 14 and 18 percent fiber. The label also gives a recommended feeding dose, which you can certainly start out with as a guideline. However, you will certainly be adjusting each rabbit's feed amount in accordance with its own personal needs.

A fiber percentage of at least 18 percent is considered to be a good choice, although a higher percentage can be good as well. Fiber is such an important part of a rabbit's diet and is therefore an important factor in your choice of pellet, particularly if you will not be supplementing with additional hay.

After settling upon a type of feed, your next question will inevitably be, how much do I feed each rabbit? This decision is best described as one part science and one part art. That is, you can start off with a formula for calculating each animal's feed ration, but you will quickly begin adjusting this base amount to compensate for many other factors, not the least of which is the rabbit's present condition.

Generally speaking, you can begin with a base feed ration of 1 ounce per pound, meaning that if you have a 5-pound Dutch rabbit, you may want to start with a base ration of 5 ounces of pellets each day. Always consider the type and amount of feed that the rabbit was previously used to eating and make your adjustments slowly and carefully over

the course of several days. If it appears that the rabbit is gaining weight too quickly and becoming too plump, you might want to reduce the base feed ration to 4 ounces. On the other hand, if the rabbit is looking a bit on the slender side, you might consider increasing the ration to 6 or 7 ounces until the rabbit has put on a bit more weight. Similarly, if your rabbit has finished eating his entire portion in a few minutes and is climbing the cage walls looking for more, you should probably consider feeding him a bit more, or at the very least giving him some additional hay to munch on.

Another guideline to go by is how quickly your rabbits eat (or don't eat) the pellets you've given them. If half of the pellets are still in the feeder when it comes time to feed again, you're probably feeding that particular rabbit too much, and you might consider reducing its daily ration until you've reached a portion that it's able to easily consume in a much shorter period of time. The exception to this guideline is that if the feed has somehow gotten wet or otherwise damaged, the rabbit will not eat it regardless of how hungry it is.

A homemade grain mixture often contains oats, sunflower seeds, and other small grains. Some breeders also include a portion of commercial pellets into the mix, but many prefer to simply feed their own concoctions of grains.

There are a couple of exceptions to the "1 ounce per pound" rule of thumb. It is usually recommended that a nursing doe should be allowed all the pellets she can eat. Again, increase the amounts a bit at a time, and wait until a few days post-kindling to begin increasing her rations. For the several weeks that she will be nursing her litter, she needs greatly increased amounts of food to keep herself in good condition and to provide milk for her growing litter. Another exception to the rule are weanling kits. You'll always have that vague worry of enteritis (digestive illness, explained in Chapter 5) looming in the back of your mind, but it's important for growing kits to have access to ample amounts of pellets. These babies grow quickly.

OTHER TYPES OF GRAINS

Some breeders prefer to create their own mixture of feed for their rabbits, rather than relying solely upon a commercial pellet. The type of grain mixture depends on each breeder's personal preferences and experiences, but it may include oats, barley, and sunflower seeds. Sometimes these grains are mixed in with a commercial pellet, and sometimes they are not. This is a more time-consuming feeding process, but some breeders strongly believe in the good results that they obtain by specializing and

One size doesn't always fit all. There may be instances when you will want to provide a couple of different types of pellets or grain mixtures to a particular rabbit. In this case, the tray on the left contains a homemade grain and pellet mixture. The tray on the right contains an 18 percent protein pellet for show rabbits.

fine-tuning their feed. I have experimented with this type of mixed feed and quickly found it to be far less efficient than feeding pellets. I had one rabbit that learned to love her sunflower seeds but didn't care for her oats. After only a few days of finding a feeder with leftover oats and all of the sunflower seeds picked out, I switched her back to pelleted feed with sunflower seeds as an added treat.

HAY

If there is one topic that we would really like to emphasize, hay would be it. Hay is surprisingly overlooked by many rabbit breeders, yet it has so many positive benefits for a rabbit. Hay is an excellent source of fiber, which is extremely important to your rabbit's diet and has been proven to reduce the risk of enteritis. Hay provides a form of roughage so that your rabbits will have the option of munching on something throughout the day, as well as a way to help relieve boredom. Rabbits love to chew, and hay provides them with an ample opportunity to chew to their heart's content.

Why is hay sometimes overlooked? Bags of pelleted feeds can be a bit misleading through their claims that the pelleted feed is a complete feed and that no supplements (including hay) are necessary for the health and well-being of a rabbit. This may be literally true, but there are additional benefits that hay can bring to a rabbit's diet, regardless of whether it is scientifically necessary.

Hay is not always readily accessible to breeders, particularly those in the city or suburban areas. Even if hay is available, it is sometimes very expensive to obtain, and it's not always easy to locate hay that has not been sprayed with harmful chemicals.

One drawback of which many breeders complain is the mess associated with feeding hay. Bits of hay dust and small stems inevitably sprinkle the rabbitry floor, and there is generally some wasted hay when rabbits are fed in the cage, although hay racks can help to cut down on spills and waste. Then there is the added time to feed hay to each rabbit at feeding time.

However, despite these drawbacks, hay has a multitude of positive benefits. This is especially true for weanling rabbits, as the increased fiber in their diets from the hay is very important in protecting

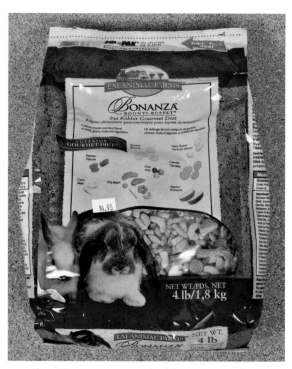

Feed mixes that are manufactured and sold in small bags can be very good for the pet owner or the small rabbitry owner, but they're not a practical option for larger rabbitries.

When shopping for hay, you may have the option of looking at different varieties, including grass hay, alfalfa hay, or a mixed hay that includes both kinds. Regardless of the type that you choose, your foremost priority is to make sure that the hay is of excellent quality and has not been damaged by rain.

Pre-packaged hay can be a good option for those who have difficulty locating hay for their rabbits. The drawback to pre-packaged hay is that it can be considerably more expensive than hay purchased in bulk directly from a farmer.

If you have difficulty in locating hay sources, another option is alfalfa cubes. These can provide additional fiber to your rabbit's diet without the hassle of trying to locate a source of hay.

them from enteritis. It's very interesting to watch the young rabbits at feeding time and to note that many of them will hop back and forth across their cage, munching a bite of hay, then a few pellets, and then some more hay. They seem to appreciate and enjoy the variety. The added fiber that is found in hay is also extremely beneficial for Angora rabbits, as it helps to keep their digestive system running smoothly and free of fur balls (wool blockages in the stomach or intestines).

Choosing Hay

Locating hay can be difficult in some areas, though it's easily accessible in the more rural parts of the United States. When selecting hay for your rabbits, one important criterion is that the hay should be free of chemicals. In addition, you will want hay that has been properly stored and protected from rain. Round bales, which are popular with horse breeders, are not a feasible choice for the rabbit breeder because their immense size makes them impossible to move without a tractor. They simply aren't practical for use with rabbits. Therefore you will be on the lookout for small square bales that

you can feed in small amounts. You will also want to purchase hay that is of good quality and meets the following criteria: a sweet, fresh smell; a greenish color (not brown, yellow, gray, or black); and low dust. Look for hay that is free of weeds, as some weeds may be harmful for your rabbits. You may occasionally find a hay bale that has a dead snake, mouse, or frog inside that was accidentally baled into the hay. Immediately discard any contaminated bales and never feed the hay to your rabbits.

Types of Hay

You essentially have a choice between two types of hay: grass and alfalfa. As with so many topics in rabbit raising, there are advocates for both types of hay. Some breeders swear by alfalfa hay and recommend it wholeheartedly; others feel that grass hay is far and away the best choice for feeding rabbits. Which piece of advice is correct?

To compare the two hay types in extremely simplistic terms, alfalfa hay is typically higher in protein and lower in fiber, and grass hay is typically higher in fiber and lower in protein. Therefore, your choice may depend slightly upon the type of pellet that you've decided to feed. If your pelleted feed is a little lower in fiber, you may want to select grass hay in order to boost your rabbits' fiber intake. On the other hand, if you're feeding a pellet that is high in fiber but is perhaps not quite as high in protein, you might want to select alfalfa hay in order to boost the amount of protein in your rabbits' diet. Your choice of hay might also depend on what is available in your area. Or you can mix the best of both worlds and look for a grass/alfalfa mix hay to enable you to feed a lovely combination of the two types.

VEGETABLES AND FRUITS

Fruits and vegetables provide variety and added nutrition to your rabbits' diet, but it's usually recommended that these special treats be fed as supplements to your rabbits' daily ration rather than the bulk of it. Always remember to introduce new foods slowly and in small amounts until your rabbits have become accustomed to the change in their daily fare.

Carrots are a universally accepted rabbit treat. Apples are also very popular with rabbits (remove the seeds), and apple wood branches are an especially satisfying treat. Dandelion leaves, which are usually available in abundance during the spring months, are also very popular with rabbits, as is lettuce (except for iceberg lettuce, which should not be fed to them). In addition to apples, other fruits that are safe for rabbits (in moderation) include raspberries, strawberries, and pears. You can brighten your rabbits' day by presenting them with one of these special treats.

Hay is beneficial for its fiber content, as well as for its use in decreasing boredom. This young Holland Lop is enjoying the contents of her hay rack.

Fruits and vegetables are a great choice for rabbit treats. You should always thoroughly wash any produce that you offer to your rabbits, since they may contain pesticides that should be removed prior to consumption.

A sugary treat, such as this yogurt drop, may be appealing to your rabbit, but it's not usually the best choice. A far better sweet treat would be a small piece of apple.

Bread is considered to be an appropriate treat for rabbits, but as you can see, bread isn't always eagerly sought by bunnies.

Salt and mineral wheels are often found in pet shops, but you must carefully decide whether to use them for your rabbits. Salt and mineral supplements are not usually a necessary addition to a rabbit's diet because commercially produced rabbit pellets typically offer ample amounts of salt to suit your rabbits' needs. These may be good supplements if you're not using pelleted feed.

Even though food is extremely important for your rabbits, water is every bit as important. Fresh, clean water should be available to your rabbits 24 hours per day, seven days a week. Your water equipment or crocks should be regularly monitored to make sure that they are full and in working order.

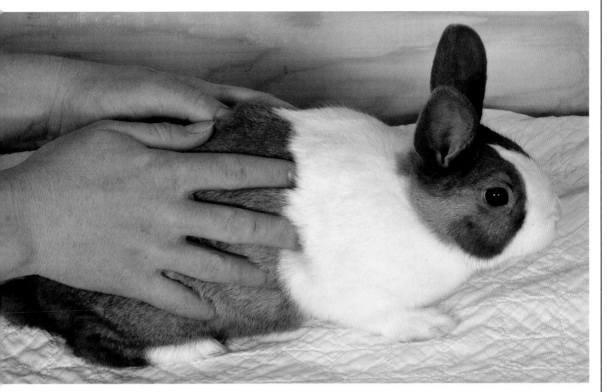

Body condition should be checked regularly on each rabbit in your rabbitry. You want to watch for rabbits that might be too thin, but you'll also want to watch out for rabbits that are overweight. Neither extreme is ideal.

TREATS

Fruits and vegetables are probably your best choice for treats for your rabbits. One other safe option would be a small slice of bread, but foods containing sugar (cookies, yogurt drops, etc.) are really not the best to feed your rabbits. The fruits and vegetables mentioned in the section above are much safer and healthier options.

WATER

We must not overlook the importance of water to your rabbits' basic nutrition and health. Water is truly the single most important thing that you can provide for your rabbits. They cannot eat without it, so always double-check your waterers at feeding time. Make sure that the bottles are filled and the nozzles are working properly, or that your water crocks are full. Refresh the water often to keep it fresh and clean, and regularly wash your water bottles or crocks to keep them sparkling clean. It's amazing how quickly water crocks can become dusty and grimy (or hairy if you have a wooly breed), so keep them cleaned regularly to encourage your rabbits to drink as much as they wish.

EVALUATING BODY CONDITION

There is a fine line between a rabbit that is overweight and a rabbit that is underweight. Neither one is ideal, although an overweight rabbit is more prone to reproductive problems. In the case of a doe, she may have trouble conceiving and/or delivering her litter if she is too plump. On the other hand, no one likes to see a thin rabbit, so you will be continually fine-tuning your feeding program to compensate for these slight changes in each rabbit's condition as time goes by. I recently purchased an eight-week-old Holland Lop that had a beautiful frame and excellent potential as a show prospect, but she seemed just a tad too thin for my preferences. After four weeks of fine-tuning her daily rations and switching her to free choice hay in addition to her pellets, she has blossomed beautifully and is in excellent condition. But when you're caring for a barn full of rabbits, it's not always easy to notice subtle changes in their body condition on a daily basis. That's why I like to take the time to regularly handle each rabbit and give it the once-over every few days or so. This way I keep up with slight changes and can quickly remedy the situation by adjusting their rations accordingly. It's not always easy to find that magical spot in between overweight and underweight, but the results are well worth the time invested.

NATURAL FEEDING OF RABBITS

Now that we have thoroughly gone over the ins and outs of pellets, grains, hay, and treats, it's time to consider an entirely different approach to feeding rabbits: the natural approach.

As we previously stated, pelleted feed has become the standard way that rabbits are fed, due to multitudes of satisfied rabbit breeders who feel that the pelleted feeds have revolutionized rabbit raising and greatly increased the production of healthier, higher-quality rabbits. The majority of rabbit breeders agree with these sentiments, but some are switching back to more old-fashioned and natural methods of rabbit feeding.

As you might suspect, natural feeding consists of natural foods, or foods that a rabbit would eat in the wild. These types of diets include plenty of fresh greens along with fruits, vegetables, and hay. Pellets are typically eliminated from the diet in a natural feeding program, although some types of natural grains may be offered, such as oats or sunflower seeds. Hay is a major staple of the natural feeding program. Even though it is not a "fresh" food, it is still a very natural one.

Of course, there are challenges to this type of feeding, especially in the winter season. In many climates, it is impossible to locate fresh greens during the winter months, and it can be expensive and time-consuming to purchase supermarket produce to feed to your rabbits during the winter. In addition, supermarket produce is often coated with pesticides, which means that you will need to wash the produce prior to feeding it to your rabbits. Another option would be to purchase organic produce, but that can increase your costs even more. Some breeders are attempting to grow their own greens indoors during the winter months by keeping pots of dandelions and growing sprouts to feed to their rabbits. This may be feasible for small rabbitries with a limited number of animals, but it would be quite an undertaking for a larger rabbitry. Most of the breeders utilizing a natural feeding program opt for one that is heavily supplemented with hay during the winter months.

Another challenge includes keeping your young rabbits at a good weight without the aid of pellets. It's not always easy to provide enough natural food to compensate for the lack of pellets in the diet of your growing juniors. Some breeders have reported problems with increased incidences of enteritis in their weanling rabbits that are kept on a natural feeding system.

Some breeders are striving for a happy medium by feeding pellets in reduced quantities and supplementing with natural foods. This type of situation can offer the best of both worlds, not to mention a savings on the monthly feed bill. There is increased time involved with preparing natural foods, however, and each breeder will have to consider the value of his or her own time in comparison to the potential benefits of a natural feeding program.

There is one significant difficulty in working on a natural feeding program for your rabbits, and that is locating natural and fresh foods during the winter months. Some breeders attempt to grow their own greens indoors during the winter months, or you can look for a product, such as the one shown here, which is a "grass cafe" that grows fresh grass in a small container. This is an interesting and innovative way to keep those greens in front of your rabbits all year round.

One of the main components of natural feeding is fresh greens. Dandelion leaves are always a favorite with rabbits. This rabbit is only receiving a small amount, which is always a wise choice when introducing a new food into a rabbit's diet.

CHAPTER 5

THE HEALTHY RABBIT

This is the kind of chapter that no one really wants to read. We have titled it "The Healthy Rabbit," but we know that you know this is the chapter in which we're going to discuss unpleasant things like rabbit ailments and illnesses, things that mar the happy planning of a new rabbitry. However, every responsible rabbit breeder needs to know the basics of health care for their herd in order to recognize when a problem arises. In this chapter we'll discuss a few of the more commonly seen rabbit ailments and give you an overview of basic health care, including grooming, exercise, and how to recognize the difference between a healthy rabbit and a sick rabbit. Best of luck to you, and may all your rabbits be healthy rabbits.

Rabbits do a very good job of grooming themselves. It's enjoyable to watch a rabbit as he cleans his face, paws, and ears.

For many years, veterinary care for rabbits was extremely difficult to locate. Today, there are more veterinarians that are trained to work with rabbits and it is easier to find a qualified veterinarian to treat your rabbits.

BASIC HEALTH CARE

We have made this statement elsewhere throughout this book, but it bears repeating again: basic health care starts with a clean rabbitry. If your cages are kept clean, the trays are emptied regularly, the cages are disinfected on occasion, your nest boxes are kept tidy, your waterers and feeders are washed regularly, and fresh water is provided at all times, the health of your rabbitry will be leaps and bounds ahead of that of another breeder who doesn't follow the same protocol of cleanliness. A healthy rabbit cannot thrive in dirty conditions. In addition, a regular feeding program is essential to the good health of your herd. This means having regular times for feeding, as well as maintaining continuity of the components of your feeding program.

There is some difference of opinion as to the best routine for feeding rabbits to promote good health. It is generally accepted that rabbits eat most readily during the evening and overnight, so many rabbit breeders give pellets and hay during the evening feeding. Some breeders feed their rabbits only once a day, while others feed pellets and hay in the morning, as well as in the evening. Others provide pellets in the evening and hay in the morning. Whichever routine you settle upon, always try to maintain regularity and feed your animals at the same time each day. Rabbits, like so many other animals, thrive on a routine and expect their food at the same time each day. Don't disappoint them.

You should never abruptly switch from one type of feed to another, and this includes pellets. Rabbits need continuity in their diets, so you will want to be very careful when making any changes.

ITEMS FOR A FIRST AID OR CARE KIT

Hopefully your first aid kit is something that you will rarely have to use. It's a good idea to have these items gathered into one location so that you can quickly access them in the event that you have a rabbit with a slight injury or illness. Always consult a veterinarian before administering any type of medication or treatment.

Preparation H (a good choice for treating sore hocks)

Cotton swabs or cotton balls

Mineral oil (for treating ear canker)

Antibiotic ointment

Scissors

Rubbing alcohol

Terramycin eye ointment

Sterile eyewash

Teeth trimmer

Nail clippers

Personally speaking, I feed my rabbits pellets twice per day. I feed hay once a day as free choice in their hay racks, and if I notice that they've eaten it all when I feed them the second round of pellets for the day, I refill the rack. I like them to have the option of munching on hay at all times. It helps to relieve boredom and keeps them busy. My twice-per-day feeding routine stems from years of raising horses and the habit of heading to the barn twice a day to feed the horses. It naturally followed that my rabbits are on the twice-daily feeding routine.

Maintaining regularity with your types of feed is as important as maintaining regularity with the timing of your feedings. In order for your rabbits to stay healthy, you must make any changes in feed very slowly. If you are switching brands of pellets, you will want to slowly add your new feed to the old feed and increase the rations of the new feed over the course of several days before entirely switching to the new brand. This is beneficial for two reasons: one, it allows your rabbits' digestive systems to become accustomed to the new pellets slowly, so that the delicate balance of their intestinal flora isn't upset; and two, it allows your rabbits to slowly become used to the taste and smell of the new feed. Many rabbits are suspicious and may refuse to eat a new type of feed that they aren't used to. For these two reasons you should always make any dietary changes slowly. Similarly, you would not want to abruptly switch hay types or suddenly introduce large portions of an unfamiliar treat, especially greens or fruits. Feeding a large handful of dandelion greens twice a week is probably not a good choice. The better idea would be to feed a dandelion green or two each day, with the smaller portions spread out over a regular feeding routine.

Abrupt changes in water can also be harmful for your rabbits. If you're traveling to shows, you might want to slowly switch your rabbits to bottled spring water a few days before you leave for a show (mix a portion of your regular water with the bottled water, slowly increasing the amount of bottled water each day until your rabbits are switched over to it), so that you can maintain the same type of water during the trip and show. Slowly switch back to your regular water by using the same method once you return home. Again, this is to avoid upsetting your rabbits' delicate digestion, and also to ensure that they will continue drinking while you travel and show. It's upsetting to reach a show only to have your rabbit refuse the taste of the strange water, so this simple precaution can help to avoid this situation.

IMMUNIZATIONS

Most types of livestock, as well as most types of pets, are typically immunized against many diseases. Rabbits are the exception to this rule, as there currently are no available vaccines for rabbits. Rabbits are remarkable healthy creatures, and the illnesses they are prone to are not ones that can be aided by the use of vaccination. Researchers have attempted to work toward a vaccination for snuffles, but as of yet this goal has not been reached.

It has been said that rabbits do not harbor any diseases that are transmissible to humans, with the exception of pinworms. On the whole, there is very little that a rabbit can do to harm a human, except maybe to scratch their arms.

PARASITES

There are several types of internal and external parasites that can affect rabbits. Internal parasites can include coccidia (a parasite of the intestines and the liver) and intestinal worms such as pinworms, tapeworms, and whipworms.

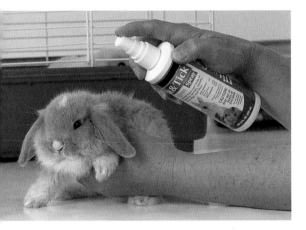

External parasites can occasionally cause trouble for rabbits, so you may need to consult a veterinarian about the possibility of using a flea and tick product on your rabbits.

A nice clean nose is just what you want to see in your rabbits. Nasal discharge can be indicative of snuffles, which is not a condition that any rabbit breeder wants to see in the herd.

Coccidiosis is not always identifiable by symptoms, and the rabbit may be infected without showing any signs. In the case of meat rabbits, coccidiosis is discovered upon butchering when white spots are observed on the liver. This causes the meat to be unsuitable for consumption. Coccidiosis can be treated or prevented through the use of a product called sulfaquinoxaline. Proper cage sanitation is also beneficial for preventing coccidiosis. Talk to your veterinarian about the possibility of occasionally treating your rabbits for internal parasites.

External parasites are more easily noticed; these include ear mites, fur mites, and fleas. Ear mites can cause ear canker (see Common Ailments section), and fur mites can cause hair loss, particularly around the head and neck. Fleas are notoriously troublesome pests that can affect your rabbits just as they affect your dogs and cats. Treatment is similar to that of a dog or cat. Your veterinarian can provide you with the best advice for treating these types of parasites.

COMMON AILMENTS

Thankfully, rabbits are generally very healthy creatures. There is a very long list of potential illnesses that might possibly crop up and make your rabbit sick, it would take a lengthy portion of this book to go over them all and would leave you in fear for all of your rabbits' lives. Therefore, we will restrict our coverage to a few of the more commonly seen illnesses that rabbit breeders should be aware of.

SNUFFLES (PASTUERELLA)

Nearly every rabbit breeder has heard of snuffles, yet not everyone understands the actual symptoms or causes. Not every snuffly-sneezy sound a rabbit makes is indicative of snuffles, as these sounds can also be caused by allergies to dust or hay. Snuffles is accompanied by thick nasal discharge, in addition to the sneezing, and is highly contagious. Some rabbits can carry snuffles without any manifested symptoms, although the symptoms usually become evident after stressful events (such as showing) or after a doe kindles her first litter. Many breeders cull for snuffles in an attempt to eradicate any carriers from their herd. Others attempt to treat snuffles with antibiotics, but generally only mixed results are obtained. Some rabbits seem essentially cured, only to have a recurrence of the disease at a later date.

ABOVE: Mastitis sometimes occurs in nursing does, so you will want to carefully observe your does for this condition, particularly right after the birth of her litter or just after weaning. An infected breast becomes hot and swollen, as well as very painful.

RIGHT: Ear canker can be treated with mineral oil or with a specifically formulated product such as this one.

Veterinarians now believe that snuffles can be temporarily placed in remission by the use of antibiotics, but the disease cannot be cured.

MASTITIS

Mastitis is an infection of the mammary glands and is seen in does that are nursing a litter, as well as does that have recently weaned a litter. Symptoms include hot, swollen teats, and treatment can include penicillin injections and hot-packs. Gradual weaning of your does' litters can help minimize the chances of mastitis, so always consider weaning the kits one-by-one to give the doe's milk supply the chance to dry up without depriving her

of her entire litter all at once. Mastitis that occurs just after kindling is more difficult to prevent.

EAR CANKER

Ear canker is a scabby type of outer-ear infection that is caused by a mite. It is not terribly common but is seen on occasion. It can be treated over the course of a few days by dripping mineral or vegetable oil in the ears to eliminate the mites. In order to prevent ear canker before it starts, some breeders treat each rabbit in the rabbitry by placing a drop of oil in each of their ears every few weeks. This heads off any potential for the development of ear canker.

Always watch for ear canker. A regular check inside each rabbit's ear will help alert you if any problem is brewing. You will be looking for any type of crustiness in the ear.

COPROPHAGY

Coprophagy is not an illness and is actually a perfectly normal behavior, but since many newcomers to rabbits are quite dismayed to see their rabbit apparently eating its own droppings, it deserves mention here. The so-called "night feces" that rabbits eat are much different than the regular droppings that you see in your rabbit's cage. Night feces are softer and look like a tiny cluster of grapes. This appearance startles new rabbit owners, and they often mistake night feces for diarrhea—which they most certainly aren't. Rabbits eat night feces in order to gain beneficial vitamins and to aid their digestion. This is a perfectly normal occurrence and it is one thing that you don't need to worry about.

SORE HOCKS

Sore hocks is a type of foot ulceration more commonly seen in the giant breeds that are housed in wire cages. It is not as commonly seen in the smaller breeds. The wounds can be treated, but prevention is also very important. It's always a good idea to provide floor mats/resting boards to give your rabbits a place to get off the wire and rest.

ENTERITIS

There are a few different types of enteritis, but one of the forms most commonly seen is mucoid enteritis, an illness that often strikes young kits from 5 to 12 weeks of age. As young rabbits mature, their diets expand from being composed of 100 percent mother's milk to include hay, pellets, and water.

The thick fur that protects a rabbit's feet and legs is very helpful in preventing certain conditions, such as sore hocks. This young Holland Lop has more than ample fur.

Wound-aide is one of the products that can be used to successfully treat certain ailments, such as sore hocks. Prevention is paramount, so always take care to prevent sore hocks before it starts.

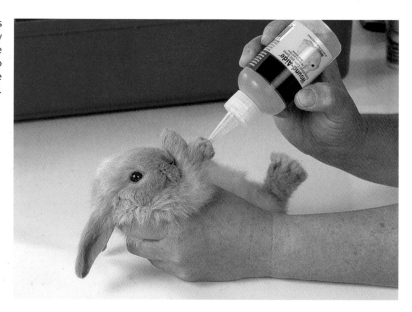

Bright, clear, and alert: This is what you're looking for in a healthy rabbit's eyes. Runny, crusty, or red eyes can be indicative of a problem.

Typically this happens very slowly to give the young kit's digestive system time to adjust and become accustomed to the new foods being introduced. However, premature introduction of green foods is believed to cause increased incidence of mucoid enteritis. The symptoms of this harmful illness include listlessness and diarrhea, and it frequently results in death. Enteritis can usually be prevented by carefully introducing new foods in small amounts and slowly increasing the amount over the course of several days. Some breeders attempt to treat enteritis through veterinary assistance, and success is achieved in some cases. Other breeders, however, believe that rabbits who have recovered from enteritis are never as hardy and strong and are better off being culled from the herd. This is a personal

decision that each breeder will have to make based upon his or her own best judgment.

SORE EYES

This is another condition for which you will want to carefully monitor your newborn kits. Their eyes usually open at around 10 to 12 days, but occasionally they fail to open due to the presence of a bacterial eye infection that causes the eyes to stick shut. If this is the case, you may want to seek veterinary assistance because the kit's eyes will need to be carefully opened and cleaned. In addition, a treatment of eye antibiotic is usually called for and is administered for several days. Sore eyes is a condition that can be prevented in some cases by keeping the nest box as sanitary as possible.

Each day (or at least every few days) you should pick up each of your rabbits and flip him or her over. If any droppings are stuck to the fur, you will want to remove them promptly.

RED URINE

Red urine is a sight that can strike momentary fear in the heart of a rabbit breeder. It is actually a very common occurrence and is rarely indicative of a serious problem. The red urine is caused by nutrients that have failed to break down entirely, thus altering the urine color from yellow to red. This phenomenon is harmless and needs no treatment. The color usually returns to yellow within a few days. If there are actual blood streaks in the rabbit's urine, you should immediately consult a veterinarian.

QUARANTINING ANIMALS

If you suspect that one of your rabbits is ill, you should immediately quarantine it to an isolated cage and always take care to feed and care for the rabbit separately so that you don't inadvertently carry any germs to your healthy animals. Similarly, it is always wise to isolate any animals that have just returned from shows, as well as any new animals that you have purchased. This will help make sure that you don't spread any illnesses that your new animals might be harboring and that your show rabbits haven't picked up any type of sickness while at the show. Isolation for several days is a small inconvenience that will reap dividends if it means that you can prevent an outbreak of any type from infecting your rabbitry.

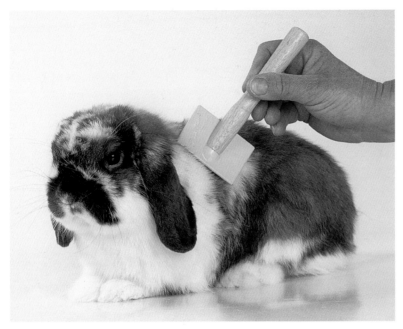

Regularly brush each of your rabbits. This helps remove any loose fur and gives you a good opportunity to thoroughly inspect each rabbit.

Grooming starts young: This two-week-old Lilac Rex is busily grooming his ears. There's nothing cuter than watching a young baby rabbit as he performs his daily "bath."

DAILY OBSERVATION

As important as clean cages, quality food, and fresh water are to the success of your rabbitry, another thing that will go a long way toward ensuring good health is daily observation. If you know your rabbits well, you will be more in tune to notice if something isn't quite right with one of them. Regularly pick each rabbit up, flip it over, and examine it on all sides. This is a good time to monitor the weight and body condition of your rabbits, as well as to check for any problems such as sore hocks, ear canker, or a runny nose. This is also a good time to check to see if any droppings are stuck to the rabbit's bottom. This problem is sometimes seen in young rabbits with long coats, such as Angoras, but is also commonly seen in Holland Lops. Any droppings that are stuck to their fur will need to be cleaned off. Recognizing your rabbit's daily habits can also be beneficial, as you'll be able to learn whether a certain behavior is normal or abnormal for your rabbit.

GROOMING

You might think that grooming your rabbits is only important if you're planning to show them or if you're raising Angoras. This is not the case. Grooming can play an important part in the care of your rabbitry whether or not you show and regardless of the breeds that you raise. Granted, your daily grooming commitment is certainly increased if you're showing and/or raising Angoras, but you should never disregard the importance of grooming on a regular basis.

Your rabbits will always benefit from a grooming session. This is particularly true when they are in the process of molting (shedding their fur).

Regardless of whether you're raising rabbits in wire cages or if you're raising them outdoors in a colony, you will still want to regularly examine and groom them.

There are a variety of brushes that are available for grooming your rabbits. The one pictured here has stiff bristles and a wooden handle.

Rabbits do not typically need to be bathed, but you can look into the possibility of using a dry shampoo that is specifically designed for rabbits, such as the one shown here.

During a grooming session, you'll be able to thoroughly examine your rabbit, check for things like ear canker and sore hocks, and evaluate your rabbit's body condition. Grooming doesn't need to be a lengthy process. It can be a quick brush or comb through your rabbit's coat, with some special attention to its belly and legs. You can also use this time to check your rabbit's toenails to see if they need a quick trim. Always be careful when trimming nails to avoid inadvertently cutting into the quick, which causes the nail to bleed profusely.

Grooming Angora rabbits can be a bit more time consuming, simply because of the vast amounts of fur. Angora rabbit owners have two methods of grooming their rabbits: the brush/comb method and the blower method. Used in conjunction, these grooming tools help keep the Angora's wool in prime condition. Angora rabbits must be groomed regularly to remove the excess loose wool. If the wool isn't brushed, the loose hair will accumulate in the rabbit's cage, and the rabbit may begin ingesting the hair while eating. This can cause a fur

Rabbit equipment suppliers and pet stores sell a multitude of rabbit products. Over time you will discover which products are beneficial and which ones aren't particularly useful for your own specific needs.

ball in the rabbit's stomach or small intestine, which can cause symptoms such as depressed appetite or diarrhea. Avoidance of fur balls is obviously the best course of action, but if you suspect that one of your rabbits is suffering from one, you can try supplementing with a small dose of mineral oil for a few days. Papaya juice or papaya tablets are also reported to be very helpful in preventing fur balls.

STOCKING YOUR GROOMING KIT

You're getting ready to show your rabbits, and you want to put together a grooming kit that contains all of the essentials that you will need to keep your rabbit in top show shape. This list will help you get started on stocking a great grooming kit.

Grooming table or stand
Grooming apron
Blower (if you're showing Angora rabbits)
Nail clippers
Combs (fine tooth, medium tooth, and flea comb)
Soft brush
Scissors
Undercoat rake
Groomer's stone (if you're showing Rex or Mini Rex)
Slicker brush
Waterless shampoo
Grooming lotion
Stain remover
Tattoo kit

EXERCISE

In large commercial rabbitries, the rabbits are rarely handled and spend most of their days relaxing in their cages. Exercise, therefore, is not an area of consideration beyond the rabbit having room to hop back and forth across his cage. Many breeders dismiss exercise as a nonessential, while others feel that it is extremely beneficial for a rabbit to have the opportunity to stretch its legs and play a bit within the confines of an enclosed area.

If you feel that exercise will be a positive benefit for your rabbits, then you might want to set up an area in which they can safely be turned loose for exercise time. This area will obviously need to be completely enclosed, and there should be no potential for a rabbit to escape. You can build a small outdoor turnout area using a wooden frame and hardware cloth, or you can purchase an outdoor "rabbit run" through pet supply stores. Always be sure that you have any doors safely latched, and always try to keep the run in a shaded area so that your rabbit does not become overheated.

MY, WHAT BIG TEETH YOU HAVE!

It sometimes surprises new rabbit owners to learn that a rabbit's teeth, like its toenails, will continue growing throughout its life. For this reason, a rabbit must have the opportunity to chew or gnaw in order to wear down the growth of the incisors. Rabbit teeth typically grow at the rate of ⅓-inch to ½-inch per month. A rabbit will occasionally suffer from malocclusion of the teeth (also known as wolf teeth or buck teeth), in which the teeth do not meet properly. This condition interferes with the proper wearing down of the rabbit's teeth as they grow. Malocclusion of the teeth is believed to be a hereditary trait, and rabbits exhibiting this defect are usually culled from a breeding program to prevent the continued recurrence of this problem in future generations. Rabbits that suffer from malocclusion may need to have their teeth trimmed by a veterinarian on a regular basis; otherwise the problem may interfere with the rabbit's ability to eat properly.

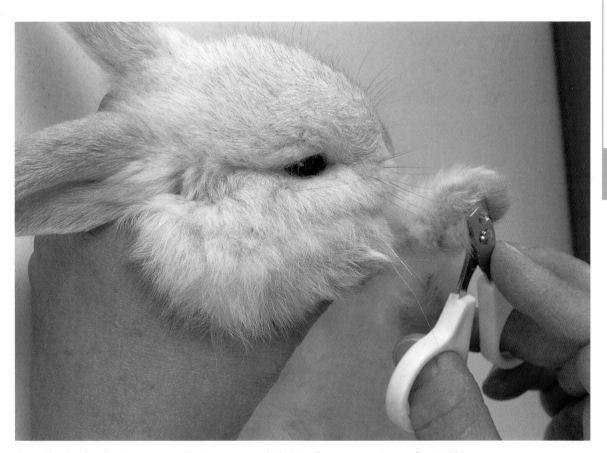

A small pair of nail trimmers or nail scissors are a vital part of your grooming or first aid kit.

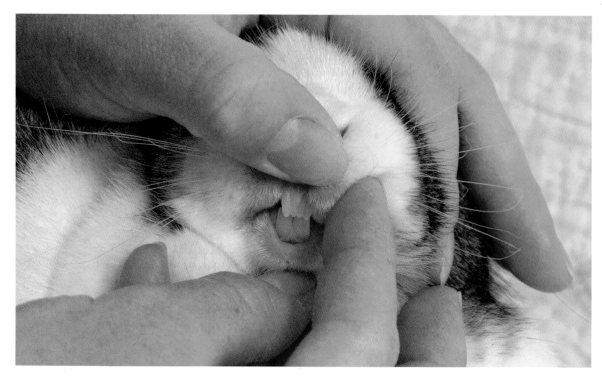

Teeth are another item that you need to regularly monitor. Every so often you should take a few moments to peek inside the mouth of each of your rabbits to examine its teeth and check for any problems. Mind your fingers when they are this close to teeth!

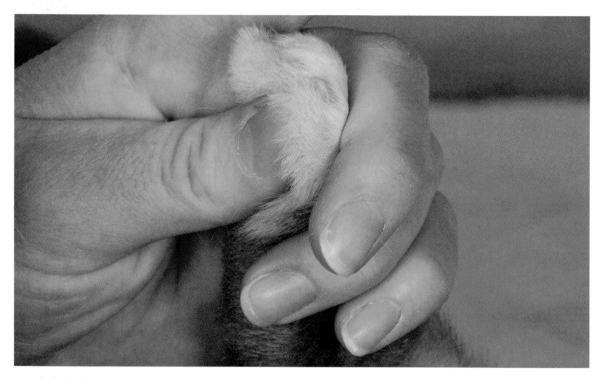

Regularly check your rabbit's feet in case its nails are in need of a trim. A special pair of nail trimmers should always be a part of your grooming or first aid kit. Keep your nail trimmers handy so that you'll have easy access to them when you need them.

Bright eyes and a healthy sheen to the coat are two of the characteristics of a healthy rabbit.

RECOGNIZING GOOD HEALTH VERSUS ILLNESS

Once you have been around your rabbits for a while, you will probably have no trouble differentiating between a healthy rabbit and an ill rabbit. Healthy rabbits are bright eyed, alert, and happy. They eat with good appetites, they have normal droppings, and they have lustrous fur. On the other hand, a rabbit that is suffering from an illness or injury is often listless and dull, has a poor appetite, may have diarrhea, and may exhibit a dullness of coat. A sick rabbit may also exhibit teeth grinding, increased respiration, and a hunched stance. As you spend time around your rabbits, you will be able to recognize the appearance of a healthy, happy rabbit and will quickly notice if something is amiss in your rabbitry.

CHAPTER 6

RAISING RABBITS FOR MEAT, FUR, OR FANCY

In all likelihood, you probably aren't looking to start up a large commercial rabbitry, but rather are seeking to develop a quality small-scale breeding program for meat, fur, or fancy purposes. Commercial rabbitries sometimes have more than one thousand breeding does in their program, which makes the majority of rabbit breeding operations look like small potatoes in comparison. We assume that most of our readers are seeking to begin breeding rabbits on a somewhat smaller scale than that example. However, some of the management techniques utilized by commercial breeding programs may also prove useful for the small-scale breeder, so let's take a quick look at some of the basics behind commercial rabbit production.

Historically, rabbits were raised mainly for meat purposes, but today the majority of rabbit breeders are raising show rabbits or breeding stock.

COMMERCIAL RABBITRIES

Commercial rabbitries sound like a place where rabbits are trained to star in TV commercials, but actually, these rabbitries simply raise rabbits on a large and impressive scale. Because of the sheer number of rabbits that are housed and cared for in a commercial rabbitry, all records must be kept meticulously in order for the rabbitry to run smoothly at all times. These records include doe hutch record cards, which contain the following information: the buck that she was bred to, the dates bred, the date due, the kindling date, how many kits were kindled, how many were alive, and how many were still alive at four weeks of age. Some breeders will record the weight of the litter at four weeks old. This information is helpful in making future decisions on whether to keep a doe in production or not. Additionally, hutch record cards for bucks can be very helpful in monitoring a buck's production as a sire. This card would include such information as the number of does bred and the number of kits per litter in order to help the breeder select for increased litter size.

A rigid schedule is also a fundamental part of commercial rabbit raising. Without a fine-tuned daily and weekly schedule, it would be far too easy for confusion and chaos to reign, particularly with regard to breeding dates, weaning dates, and breed-back dates. On top of this, many commercial rabbitries work with a complicated system of line-breeding in order to produce the best quality and characteristics possible.

Litter size is one of the main characteristics that commercial rabbitries breed for, which is quite understandable. When you are spending the same amount of time, effort, and money on a doe that produces an average of four kits as opposed to a doe that produces an average of eight kits, it's obvious which is the more valuable animal from a commercial standpoint. Her daughters would also be more valuable for breeding purposes if they too are genetically predisposed to producing higher numbers of kits per litter.

Commercial breeders are also select-breeding for does that have ten teats, as opposed to the typical eight teats. Research has proven that does with a higher number of teats regularly produce higher numbers of kits per litter than does with fewer teats. Therefore, in light of the fact that larger litters are the aim of commercial breeders, this research may prove very beneficial for breeders when selecting replacement does or initially establishing their rabbitry.

The single most popular breed chosen by commercial rabbit breeders, the New Zealand White is chosen for its feed-to-meat conversion ratio and its white coat.

Along with the New Zealand White, the Californian is also a very popular commercial breed. Even though the ears and nose of the Californian are dark, this does not affect its usefulness as a meat rabbit, even with producers who will only accept white rabbits.

Slightly smaller, but still very useful, the Florida White is another good option for those who are looking to raise rabbits for meat.

Litter size is a very important quality that is sought by commercial rabbit breeders. Obviously, the larger litters are more beneficial to a commercial rabbitry than the smaller litters, so brood does are carefully chosen for their ability to produce large litters. This doe, with her tiny litter of two, would not be considered a good choice for a commercial breeding program.

Conception rate is also a consideration for commercial breeders. Does with problematic reproductive systems (though rare, they do occur; I have one in my barn right now) are not good choices for a commercial breeding program. Neither are does that consistently have problems rearing a litter, whether this is due to issues with scattering her kits, cannibalism, or any other issue.

HOW MANY LITTERS?

A small-scale meat producer or a fancy rabbit enthusiast might aim for three or four litters per year per doe, but larger rabbitries often attempt to raise seven or eight litters from each doe per year. This requires that the does be bred back at 14 or 21 days after kindling, which further requires that her current litter be weaned at approximately four weeks of age to facilitate this schedule. Top-quality nutrition of the doe is an absolute necessity under this type of breeding schedule. Many breeders believe that a feed that is higher in protein (17 or 18 percent) is beneficial to breeding does, as well as to young fryers. "Fryer" is the term commonly used to describe a young rabbit that is being raised for meat purposes; it generally refers to an animal that is under 5 pounds and under 10 weeks of age.

WHICH BREEDS ARE BEST?

The vast majority of commercial rabbit breeders raise either New Zealand White or Californian rabbits for several reasons. Their size is suitable for reaching the ideal market weight of 4.5 to 5 pounds by the time the fryer is eight weeks old. In addition, their white coloring is generally preferred by processors over rabbits with colored pelts. Florida Whites are also used regularly in commercial programs because of their white pelts, but because they are smaller than New Zealands and Californians, they do not always reach the ideal market weight as quickly as the larger breeds.

MAKING "CENTS" OF COMMERCIAL FEEDING

When it comes right down to it, the actual profit made off of a single fryer is usually less than one dollar. However, one must evaluate the bigger picture, which includes the sale of hundreds or even thousands of fryers per year. This profit margin is greatly influenced by what is known as the "feed-to-meat conversion ratio." The conversion ratio is based upon the amount of feed necessary to produce a fryer of market weight. A good ratio is considered to be 4:1, that is, 4 pounds of feed used to produce 1 pound of meat. This is where the use of top-quality feed becomes so important. If it takes five or six pounds of a lesser-quality feed to produce one pound of meat, your feed-to-meat conversion ratio becomes 5:1 or 6:1, which significantly drops your profits.

ON A SMALLER SCALE

You probably aren't seeking to begin a rabbitry with one thousand does, but you can certainly glean some good tips from the experiences of breeders who produce large numbers of rabbits each year. Perhaps you will want to pay even closer

Commercial producers of rabbits often send their rabbits to a meat processor, where the meat is processed into products, such as the ones seen here.

With the current resurgence of interest in locally produced meat, the type of rabbit meat seen here appeals to some consumers. The designation of "no other meats added" helps to clarify that the product is 100 percent rabbit meat.

consideration to the does that you select for your program and choose for such characteristics as litter size and conception rate in addition to the other qualities that you're already seeking (breed type, markings, good mothering skills). It's always wise to consider the advice and knowledge of other breeders, especially when you're starting out.

MEAT PRODUCTION FOR HOME USE

In addition to the commercial rabbit breeders who raise meat rabbits for sale, there are many other rabbit breeders who raise meat rabbits for their own use. Today, with the increased interest in the "locavore" movement of purchasing locally and organically grown food, many people are looking at rabbits with renewed interest. This is why many hobby farmers are considering the possibility of raising their own meat and are joining the increasing number of farmers raising rabbits.

In these instances, the choice of breed is less important than it is when establishing a large commercial rabbitry. As we stated, commercial rabbitries typically work only with Californians, New Zealand Whites, and sometimes Florida Whites, due to the demand for a white-pelted meat in the rabbit meat industry. When you are raising rabbits for home use, you are not under any of these restrictions. Many of the other large rabbit breeds can be excellent choices for meat production, including the Champagne d'Argent, the Satin, the American Chinchilla, the Palomino,

the American, the Beveren, the Cinnamon, and the Silver Fox.

Breeders who aim to produce meat for their own family are not bound to the same tight scheduling that commercial rabbitries must adhere to in order to maximize their production. Perhaps three litters per doe per year are enough to fulfill your family's needs, rather than five or six litters. In these instances, you can take extra time to wean the litters or allow the fryers an extra week or two to gain weight if necessary. When you aren't bound by any timetable, the entire process of raising meat rabbits can be a much more relaxed experience.

RAISING RABBITS FOR WOOL

Riding on the coattails of the increased interest in Angora sheep and goats is a growing interest in Angora rabbits. Although there is little large-scale production of Angora rabbit wool, many small-scale breeders market this special product for hand-spinning and weaving.

To begin any type of breeding program for Angora wool, you will need Angora rabbits. The English Angora and the French Angora are two very popular types, and their wool may be gathered by plucking. Satin Angoras, which have that lustrous combination of Angora wool with the transparency of the Satin coat, can also be plucked, but the Giant Angora must be shorn. It goes without saying that selection for wool quality will be paramount when choosing your foundation stock.

The Angora breeds are chosen by those who are interested in raising rabbits for wool purposes. The French, Satin, and Giant Angoras are all commonly used, as is the English Angora, such as the one seen here.

Wool quality is paramount when raising rabbits for wool purposes, so always carefully evaluate the characteristics of the wool in your breeding stock.

Due to their smaller size, Jersey Wooly rabbits cannot produce the volume of wool that some of the larger Angora rabbits can; however, some breeders prefer to work with these smaller wooled rabbits. The Jersey Wooly was developed through crosses of Angora rabbits on Netherland Dwarf rabbits. This Jersey Wooly exhibits the large eyes for which the Netherland Dwarf is noted.

MARKETING YOUR RABBIT WOOL

Marketing rabbit wool can be an interesting and unique way to supplement the income from your rabbitry, and it helps if you have a personal interest in hand-spinning and working with Angora wool. The key to marketing your wool or wool products is to let people know what you have available. This requires exposure. Craft shows are an excellent place to start. Perhaps you can purchase a booth at a well-attended craft fair and sell your wool and wool products while doing a demonstration for passersby. Or perhaps you can consign your items to local gift shops or take them to flea markets. The Internet offers a wide variety of opportunities to market your products. If you aren't ready to set up your own website, you can try marketing your products on www.ebay.com or through an Etsy shop (www.etsy.com). Contact your local newspaper to see if they can arrange a small press release explaining your products and your work with Angora rabbits. The possibilities are endless.

In some breeds, the pattern of white markings is an extremely important part of their breed standard. These Dutch rabbits are a good example of the variation that can be found in rabbit blazes. The doe on the left has a narrower blaze than the doe on the right, whose blaze is more closely matched to the *Standard of Perfection*. These are important characteristics to consider when raising fancy rabbits.

Jersey Woolies and American Fuzzy Lops also exhibit wool coats but are less commonly used by breeders who are raising rabbits for wool. The larger breeds typically produce a higher yield of wool due to their larger size.

RAISING FANCY RABBITS AND BREEDING STOCK

By far, the majority of people who raise rabbits are doing so for show or breeding purposes. They outnumber the breeders who specialize in meat production or specialize in fur. The intrigue of raising fancy rabbits (i.e., show prospects) appeals to breeders of all ages and from all walks of life, as evidenced by the incredible support that is found at rabbit shows nationwide. The fascination extends much farther than the show table. Breeders find great satisfaction in attempting to achieve various goals: the perfectly marked specimen, the precise shade of red on a Thrianta, or the proper crown on a Holland Lop. The pursuit of these ideal characteristics, as listed in the *Standard of Perfection*, are part of the thrill that keeps rabbit breeders involved and enthused about their breeding programs.

Some breeds are raised exclusively for fancy purposes. These breeds would include any of the smaller breeds that do not have potential for any other purpose (meat or fur), such as Netherland Dwarf, Britannia Petite, Polish, Holland Lop, and Dwarf Hotot.

After you've been raising rabbits for some time, you will ideally have more quality rabbits than you're able to retain for your own breeding program; at this point, you may be able to begin marketing some of them as breeding or show stock. If you have six gray Dutch junior does in your barn, chances are that you won't need to keep all of them as breeding or show stock for yourself. In this case, you might keep back two or three of them that you feel have the quality and specific characteristics necessary to further improve your breeding program and offer the others for sale.

Sales of show- and brood-quality rabbits can be a more lucrative way to earn income from your rabbitry than by working with meat or fur. Prices for purebred rabbits of show or breeding quality can range anywhere from $10 to $350, with most landing somewhere in the $35 to $100 range for quality junior rabbits. The vast majority of rabbit breeders are involved with raising rabbits strictly as a hobby, but the sales of rabbits can certainly help to offset some of the costs of feed and equipment.

SELLING SHOW-QUALITY RABBITS

As we'll discuss in Chapter 10, showing your rabbits can be extremely beneficial in order to validate the quality of your breeding program, but selling show-quality rabbits is another way to truly increase exposure for your rabbits. If the new owners of your rabbits are showing rabbits from your breeding program (and better yet, doing well on the show tables with them), this is a very valuable promotion for your rabbitry.

Selling a rabbit as "show quality" generally means that you feel that the rabbit has the potential to do well on the show table, and more precisely, that the rabbit has no known problems or disqualifications that could prevent it from showing successfully. Show quality also means that the rabbit is an acceptable color, appropriately marked for its breed, and the proper weight. Some sellers offer a guarantee that if they sell a rabbit as show quality and it is later disqualified (DQ'd) from competition,

they will substitute a different rabbit or refund the purchase price.

SELLING BROOD-QUALITY RABBITS

Selling brood rabbits is not quite as cut-and-dried as selling show stock. With breeding stock, there is sometimes that gray area of deciding whether a particular fault or a particular disqualification (DQ) is sufficient to make the rabbit unsellable as a breeding animal. For instance, I don't like my Dutch to have long or split stops. This is important to me. I want their leg markings to be correct, but that's my personal choice. I know that I'm probably missing out on some top-notch Dutch breeding prospects because I won't consider any with long or split stops, but I also know that I'm concentrating my bloodlines on properly marked specimens, and that should (hopefully) increase my percentages of properly marked offspring. On the other hand, there are many very successful Dutch breeders who wouldn't hesitate to purchase and breed from an animal that had a split or long stop, as long as the rabbit's other characteristics were overwhelmingly positive.

Like the blaze, the saddle (a white marking around the shoulder and neck) is an important characteristic of a Dutch rabbit. As you can see, the rabbit on the left has a slightly jagged saddle. This could certainly affect her performance on the show table, even though many of her other characteristics are quite admirable. The doe on the right has a more correctly placed saddle, but she is not quite as nice in body type as the doe on the left. These are all areas that you will need to consider when deciding which rabbits should be sold as brood prospects and which ones merit being sold as show prospects.

Within the image, the following text appears on the open book:

DUTCH
VARIETIES: BLACK--BLUE--CHOCOLATE
GRAY--STEEL--TORTOISE

SCHEDULE OF POINTS

GENERAL TYPE		
Body	17	25
Head	5	
Ears	2	
Eyes	1	
FUR		10
COLOR		10

MARKINGS		50
Cheeks	12	
Blaze	5	
Neck	5	
Saddle	10	
Undercut	8	
Stops	10	
CONDITION		5
TOTAL POINTS		100

SHOWROOM CLASSES & WEIGHTS

Senior Bucks & Does—6 months of age or over, weight 3½ to 5½ pounds. Ideal weight 4 ½ pounds.

Junior Bucks & Does—Under 6 months of age. Minimum weight 1½ pounds.

NOTE: Juniors may be shown in a higher age classification. No animal may be shown in a lower age classification than its true age.

GENERAL TYPE

BODY—Points 17: Dutch type consists of a nicely rounded back, beginning immediately behind the head, progressing over the shoulders to the highest point at the loin and hips (perpendicular to the stifle), and rounding off into a well filled hindquarters. From the top view, the shoulders should be rounded and slightly narrower than the hips. The rib spread should be so that the side lines taper from the hips to the shoulders. The hips should have a well rounded appearance and be full to the base of the hindquarters. The Dutch should maintain a close coupled, well rounded appearance, whether it weighs 3½ or 5½ pounds.

105

It's not always easy to determine whether a rabbit is show quality, brood quality, or pet quality. Physically comparing your rabbits to the *Standard of Perfection* can be a good first step, as the description can be evaluated point-by-point to see how each rabbit measures up.

If raising fancy rabbits catches your fancy, you might want to consider raising one of the extremely popular breeds, such as the Holland Lop, the Netherland Dwarf, or the Mini Rex.

Because of personal preferences, it's difficult to tell people that they should or shouldn't use an animal for their breeding program. Obviously, if a rabbit has a truly blatant fault that anyone would agree should not be bred from and pass along its faulty genes, that's one thing, but each breeder has to use his or her own best judgment when it comes to animals that have a somewhat subjective fault, such as weak shoulders or light bone. As long as the buyer has an ample opportunity to evaluate the rabbit and its characteristics, the rabbit's potentials and weaknesses can be sized up quite easily. In the second place, as long as you have mentioned the rabbit's faults, it's up to the buyer to decide whether to go ahead with the purchase. You may be focusing only on the rabbit's faults, but your buyer might be enamored by the positive qualities that the rabbit possesses—positive qualities that might be exactly what the buyer's current program is lacking.

You would never want to sell an animal for breeding stock if it possesses any type of hereditary defect. This rabbit would need to be either culled as a meat or pet rabbit.

Sometimes you will have a very nice rabbit of good quality that might be a bit over the standard senior weight for its breed. These rabbits are sometimes sold as breeding stock, as knowledgeable breeders occasionally use a rabbit of slightly larger size to infuse some new genetics into their breeding program—particularly if they are currently producing rabbits that are smaller than desired.

It's an unspoken rule of thumb, but if a rabbit is being marketed as a brood-type rabbit, this generally means that while it is suitable for use as a breeding animal, there is probably some reason that it is unsuitable for showing. Sometimes this is as simple as size or a slight mismark, but it is something to consider. A show-quality rabbit certainly has the potential to be a breeding animal later on, but a brood-quality rabbit is usually not going to be a show specimen.

FINDING YOUR NICHE

Although it's not imperative that you find a niche to specialize in, you may find that it will help you establish yourself as a source for fancy rabbits or breeding stock if you specialize in a particular type or variety. This may be something as broad as specializing in rare breed rabbits and establishing yourself as a source for Americans, Beverens, Silver Foxes, and Belgian Hares. Or you may settle on a niche that is a bit more specialized, such as Blue Otter Netherland Dwarfs. Once you've settled upon your area of specialization, you can really focus on doing your best to produce the highest-quality rabbits that you can. Sometimes this is helpful because you can narrow down your distractions and truly work toward perfection in your breeding program. If you're raising Rex rabbits in all colors, you might feel a bit overwhelmed by the possibilities and potential. If you're working toward breeding a specific variety, such as the Lilac color, you can begin to really focus on raising the best Lilac Rex rabbits that you can produce without the added distractions of trying to perfect rabbits of a dozen additional colors.

An area of specialization can also help when you're marketing your rabbits, as it gives potential clients a distinctive attribute to remember about your rabbitry, which can boost your name recognition among other rabbit enthusiasts. For instance, thousands of people raise Holland Lop rabbits, but far fewer specialize in the Blue-Eyed White variety. Finding your niche can go a long way toward making you a "big fish in a small pond," (i.e., a top breeder of a variety that only a few people work with) as opposed to a "small fish in a big pond" (i.e., a good breeder of a popular variety who must try hard among stiff competition).

Many rabbit breeders specialize in a particular color variety and focus their energy toward producing the most perfect specimens possible.

It often takes years to develop a new breed or color variety, but despite the potential challenges and the patience required, it's an utterly rewarding pursuit.

DEVELOPING A NEW BREED OR VARIETY

This is one of the topics that is near and dear to the heart of many rabbit breeders: the development of new rabbit breeds and/or varieties. The only reason there are forty-seven ARBA-recognized rabbit breeds is that many breeders over the course of many years have worked hard toward the development of these individual breeds. The process of achieving breed status is a lengthy and complicated journey, one that requires extensive commitment, determination, and passion.

The challenge is a part of the fascination and thrill of working toward the development of a new breed. If you talk to any long-term breeder who has attempted or succeeded in such a pursuit, you're bound to catch a taste of the drive and determination that keeps them going even when discouraging things happen and setbacks occur. The joy of working toward something as fulfilling and rewarding as developing a new breed simply outshines the difficulties.

On a smaller scale, the development of new varieties can be every bit as rewarding and every bit as challenging. As of this writing, there are numerous breeders who are working toward the development of new colors in the Mini Satin breed. Each is working steadily toward furthering his or her chosen color through presentations at ARBA conventions. The development of new varieties is a very worthwhile pursuit and one that interests many breeders.

SELLING TO THE PET MARKET

Every breeder that raises show- or breeding-quality rabbits will have the occasional cull (mismarked kits, kits that don't exhibit good breed type, or kits with disqualifying factors). This is an inevitable part of rabbit breeding. Many breeders use these culls themselves for meat purposes and feel that they gain the best of both worlds: the opportunity to raise top-quality rabbits of a specific breed while still obtaining the benefits of homegrown meat. However, this is not everyone's cup of tea, and there are some rabbit breeders who won't consider this option. Instead, these breeders market their culls as pets. In addition, some breeders are now working specifically to breed for the pet market. If this is something that interests you, the best advice is to thoroughly consider your market before you begin breeding. Do you have a good outlet with which to market your pet rabbits? Can you enlist the help of a local pet shop? Do your homework ahead of time to determine the nature of your market to make certain that there will be sufficient interest in your pet rabbits.

One rule of thumb to remember is pet rabbits should be cute. The majority of people purchasing a rabbit as a pet will be looking for something that is adorable to look at, and the cute factor will help to increase your sales. Another consideration is size. Large rabbits can be too cumbersome for a young child (your number-one pet customer) to handle, yet the smallest breeds of rabbits (Netherland Dwarf, Britannia Petite, Polish) have gained the reputation of being too high-strung to be children's pets. Therefore, you will probably want to consider a breed that is in the 4- or 5-pound range, such as the Mini Satin, the Mini Rex, or the Dutch. All of these have good reputations for pleasant dispositions and definitely exemplify the word "cute!"

Although there is a ready market for pet rabbit sales around Easter, some breeders are leery of making sales around that holiday. The gift of an Easter bunny is often fun for a few days, but children often tire of caring for their pet and many Easter bunnies are quickly forgotten.

A pet-quality rabbit doesn't necessarily indicate a rabbit of inferior quality. This young girl has two very nice Dutch rabbits as her pets, yet they both retain the quality necessary to make good breeding animals if she decides to raise rabbits in the future.

PET-QUALITY VERSUS SHOW-QUALITY

If you are raising rabbits for show or breeding, let's imagine that your first few litters have arrived, and your charming new kits have just emerged from their nest boxes. They are all so cute and adorable, and their antics are delightful, but you may be struggling with a small problem. How can you determine which ones are show-quality and which ones aren't?

First of all, let's make it very clear that they won't all be show-quality, unless you've established some kind of magical breeding program that the rest of us haven't been able to create. It's more than likely that in each and every litter there will be above-average kits, average kits, and below-average kits. As your breeding program progresses and hopefully you're able to work toward higher and higher quality, your number of above-average babies will increase while your number of below-average babies will decrease. But it's a fact of life that not every kit is going to be a show stopper. The trouble comes from telling the difference.

Second of all, let's make it very clear that you shouldn't expect to be able to differentiate between show-quality kits and pet-quality kits when you're first starting out. It takes time, study, education, and experience before you will be able to look confidently at a litter of rabbits and be able to evaluate them. Even once you've raised several litters and had time to watch the youngsters grow into mature rabbits, you still might struggle when your junior rabbits go through a stage and you don't know for certain whether they're going to grow out of it. Even long-time rabbit breeders admit that it isn't always easy to determine which rabbits are going to be the pick of the litter.

What if you have a litter of eight-week-old kits and aren't sure which ones to keep as show and breeding prospects or which ones to part with as pets? You might just want to take your time and wait. Let them grow up a bit more, let them mature, and keep watching them. You can always sell the pet-quality rabbits later. It's not worth the risk of inadvertently selling a top-notch show rabbit before you can tell for certain.

Even though this young boy and this dwarf rabbit make an adorable pair, dwarf rabbits are generally considered to be an unwise choice as a child's pet. Dwarf rabbits have a reputation for being a bit more high-strung than some of the other breeds, but this is sometimes an unfair generalization. There are certainly many dwarf rabbits with excellent dispositions that make wonderful children's pets.

You can also enlist the help and assistance of a long-time rabbit breeder, ideally one who has experience with your breed. Ask the breeder to come over to evaluate your juniors and give you an opinion on the quality and potential of each rabbit. This kind of assistance can go a long way toward helping you to gain an understanding of young rabbits, their rate of maturity, and the characteristics that you should be looking for when evaluating future litters.

OTHER COMMERCIAL ENTERPRISES WITH RABBITS

There are a couple of additional ways that you can boost your income from your rabbitry, both related

to something that your rabbitry will undoubtedly have in abundance: rabbit manure.

RAISING WORMS

If you are raising rabbits, you can easily raise something else, too. Many rabbit breeders take advantage of their endless source of manure/ fertilizer to raise earthworms. As unlikely as it might sound, many breeders are able to supplement their rabbit income by raising earthworms directly underneath their rabbit cages in rabbitries where there are no trays underneath the cages and the droppings fall directly to the ground. In the soil beneath the rabbit cages, worms can be easily raised, gathered, and sold as fishing bait. The process of raising worms is really quite simple and straightforward. Once you're set up and ready to go, it takes only a bit of time each week to rake the earthworm beds underneath the cages. With only a few worms to start with, you can raise literally thousands of worms in a relatively small rabbitry. There's a very handy book available on the subject of raising worms titled *Raising Earthworms for Profit* by Earl B. Shields.

MANURE/COMPOST

Perhaps raising worms isn't exactly your cup of tea, or you just don't have the time to gather night crawlers and collect them for sale. You still have another option for making your mountains of rabbit manure work for you, as compost.

Rabbit manure has a wonderful reputation for being the ideal foundation for compost, and its benefits are well-regarded by gardeners. There are wonderful books available on the subject of making

compost. The process is too lengthy to cover in-depth within the confines of this chapter, but let's take a quick look at the benefits of making compost from your rabbit manure.

The first major benefit is that making compost provides a positive use for the continually expanding pile of manure that your rabbits are contributing to on a daily basis. Secondly, creating compost is an environmentally friendly pursuit, as it combines the recycling of manure with the recycling of other discarded items (kitchen garbage, leaves, etc.) and creates a new and useful product that is of great value to gardeners.

The third benefit is that you might be able to supplement your rabbitry's income by selling your compost to local gardeners or landscaping professionals. At the very least, you should have no trouble locating individuals who are more than eager to haul away your compost (or even your rabbit manure, if you aren't inclined to make your own compost) for free.

As a breeder, if you want to provide rabbits as pets for children, one of your first priorities will be to select only the rabbits with the finest temperaments as potential children's pets.

CHAPTER 7
YOUR PET RABBIT

Throughout this book, we've fully discussed the ins and outs of raising rabbits commercially, raising rabbits for show, and raising rabbits for breeding stock, but it's time to take a moment to discuss another aspect of rabbit ownership that we have thus far overlooked—the joy of rabbits as pets.

Rabbit breeders raise rabbits because they enjoy doing so and enjoy the rewards and benefits of their hobby. Yet many people enjoy rabbits but don't have the desire to raise them on any scale; they choose to have rabbits as pets only.

This is where there can be a bit of confusion. People who raise rabbits for showing or breeding purposes typically have three categories of rabbits for sale: show rabbits, brood rabbits, and pet rabbits. In the majority of cases, the pet rabbits are those that did not meet the ideal criteria for some reason that prevented them from being viewed as show specimens or brood stock. They may have a disqualifying attribute, be oversized, or be mismarked. In any case, pet rabbits typically have some characteristic that prevents them from being an ideal specimen. In many cases, people who purchase rabbits as pets are more concerned with the rabbit's disposition than they are with the rabbit's breed type. However, not everyone who wants to own a few pet rabbits wants to own rabbits that have blatant faults. Therefore, anyone shopping for pet rabbits should clearly differentiate whether they want pet-quality rabbits or show-quality rabbits to be enjoyed as pets.

Rabbits make excellent pets. They have many positive attributes that make them an ideal choice for many situations when other, more "traditional" pets are not an option. Rabbits are quiet and don't require long walks or other forms of daily exercise, they don't eat very much, and they certainly don't bark at the mailman. However, a pet of any type is a definite commitment, and a rabbit is no exception. Because a domestic rabbit's life expectancy is approximately 7 to 10 years, a pet owner is committing to a long-term period of ownership and dedication, but also a long-term period of enjoyment from the pet.

Rabbits are an ideal project for children who are involved with 4-H. Children are able to get acquainted with the responsibilities of owning and raising a livestock breed while gaining the benefits of working with a small animal that is easily handled and cared for.

There are different schools of thought regarding the topic of whether bucks or does make better pets. Generally speaking, it's acknowledged that a neutered male is probably the quietest and most docile choice for a pet. However, many people have had excellent success with pets of either gender, and that includes unaltered bucks. If I were choosing a pet, I would choose a doe, but that's my personal preference. I like the fun dispositions of does, although there are certainly cases in which a doe can be unpredictable or moody. I have one doe that will attack the nozzle of her water bottle if the water doesn't come out quickly enough to suit her. She is a sharp contrast to another of my does, who would never dream of being anything but sweet and kind. There are obviously personality extremes in either gender, and it really comes down to individual rabbits and their personalities, rather than a cut-and-dried case of one gender over another.

Children love rabbits, and a rabbit can make an excellent pet and/or 4-H project. Their small size makes them easily handled by children, giving them the experience of working with animals. Rabbits are generally hardy and easy to care for.

Daily observation of your rabbits should be an important part of your daily routine. In addition to the obvious benefits to your rabbits' health and well-being, such observation allows you to truly enjoy your rabbits, thereby enriching your daily chores.

It is fascinating to observe the inherent personalities of animals, and rabbits are particularly entertaining with their mannerisms and antics. Personality is a facet of rabbit ownership that has the potential to be overlooked when rabbits are raised for commercial purposes, but pet owners have the opportunity to really get to know their rabbit(s) and observe the intricacies of each personality.

Just as with humans, there are shy rabbits and rabbits that are more outgoing and bold.

I remember bringing home a pair of 10-week-old does who had been raised together and handled exactly the same amount. One doe was always happy to see you, always climbing the hutch door, quite fearless and bold. The other doe would hide her face in the corner of the hutch whenever you spoke to her; she was quite skittish and shy. There was no explanation for the vast differences in their behavior except for inherent disposition.

Keeping rabbits as pets will give you and your family the opportunity for hours of fun watching the delightful and endless antics of your animals.

Climbing is a good form of exercise for rabbits. This small wooden platform allows this Rex rabbit to use her muscles and move about to give her a good opportunity to obtain some exercise that she might not be able to get if she were in an all-wire cage.

Rabbits are inherently curious and love to explore. This makes them particularly endearing as pets, but it also means that you need to carefully monitor them when and if you decide to let them play inside your home.

HOUSE RABBITS

Generally speaking, the term "house rabbit" doesn't necessarily refer to a rabbit whose hutch is kept indoors, but rather to a rabbit that is allowed to roam freely throughout a house as a pet. People enjoy this type of rabbit ownership because it allows them to truly get to know the natural actions and character of their pet rabbit, due to the ample observation time that comes from having a pet whose habitat is your living room.

This electrical cord is an accident waiting to happen. Always thoroughly rabbit-proof your home before allowing a rabbit to roam freely indoors. Rabbits are notorious chewers, therefore teeth and electrical cords are not a good mix.

PLANTS THAT ARE POISONOUS TO RABBITS

This list is by no means exhaustive, as it would be impossible to list each and every plant that might prove harmful to a rabbit. Always keep plants out of the reach of your rabbits unless you are positively certain that they are not harmful for rabbits to ingest.

Aloe Vera
Amaryllis
Anemone
Asparagus Fern
Autumn Crocus
Azalea
Begonia
Bittersweet
Black Nightshade
Bleeding Heart
Boxwood
Buttercup
Caladium
Calendula
Carnation
Ceriman
Christmas Rose
Clematis
Crown of Thorns
Cyclamen
Daffodil
Daisy
Deadly Nightshade
Delphinium
Dieffenbachia
Dracaena
Elephant Ear
Foxglove
Geranium
Gladiola
Glory Lily
Hemlock
*Poison
and Water
varieties*

Holly
Hyacinth
Hydrangea
Impatiens
Iris
Ivy
*Boston
and English
varieties*
Jerusalem Cherry
Juniper
Laburnum
Larkspur
Lily of the Valley
Lobelia
Lords and Ladies

Marsh Marigold
Meadow Saffron
Mexican Breadfruit
Monkshood
Morning Glory
Mother-in-Law
Narcissus
Oleander
Onion
Parsnip
Peony
Pencil Cactus
Philodendron
Poinsettia
Poppy
Potato

Primrose
Privet
Rhododendron
Rhubarb
Schefflera
Snow-on-the-Mountain
Solomon's Seal
Sweet Pea
Swiss Cheese Plant
Tomato
Tulip
Umbrella Plant
Violet
Wisteria
Yellow Jasmine
Yew

Keep out of the reach of . . . rabbits? This handy list will help you to avoid many plants that could pose a potential hazard to your rabbit.

A house rabbit can provide hours of enjoyment to a small child because it allows them the opportunity to spend a great deal of time interacting with their pet, which is interaction that might not occur if the rabbit were housed in a cage for the majority of the day.

House rabbits can benefit from the increased amount of exercise and playtime that caged rabbits usually don't get. They seem to love the freedom and will entertain you with their leaps and bounds as they run, jump, and play.

Never leave your rabbits unattended on any type of furniture, even for a moment. A potential fall or jump from a chair or sofa could seriously injure your rabbit.

RABBIT-PROOFING YOUR HOME

Because of the dangers that a rabbit may encounter when roaming loose in a house, it is quite common for rabbit owners to restrict their house rabbits to only a room or two. In this case, the room(s) should be thoroughly "rabbit proofed" to eliminate any sources of potential danger or harm that the

rabbit might encounter. Rabbits are naturally prone to chewing and may decide to chew on something that could harm them, such as an electrical cord. On the other hand, they might also decide to chew on something that you might rather they did not, such as an expensive sofa or end table. For these reasons, you will want to carefully select the objects that are accessible to any rabbit that is allowed to roam freely in your home. This is for your own benefit, as well as theirs.

It is important to remove any houseplants when you are rabbit-proofing. At best, you wouldn't want your rabbit to gorge itself on greens, and at worst, many houseplants are poisonous to rabbits.

House rabbits must be trained to use the litter box, which is not a difficult task but is certainly something that you will want to take care of before your rabbit is allowed to roam around.

Although it's a topic that is often overlooked by many rabbit owners, toys can be thoroughly enjoyed by rabbits. If you wish to provide a toy or two, your rabbits will undoubtedly put them to good use. Toys help keep rabbits entertained and they help to relieve boredom.

Don't be surprised if your house rabbit occasionally damages an object in your home. If you've carefully rabbit-proofed the room or rooms that your rabbit inhabits, then you can certainly minimize the possibility of damage, but they still might surprise you by chewing on something that you hadn't expected.

Outdoor exercise is always a pleasant opportunity for a rabbit. Even if your bunny is a house rabbit, you might want to have an outdoor run available so that he or she can enjoy the fresh air and sunshine.

It's important to select healthy treats for your rabbits and to only offer them in moderation.

TREATS FOR RABBITS

All pet owners like to make their pets happy, and one of the most common ways to do this is to provide an occasional treat for their pet. Rabbit owners are no exception, and many enjoy giving treats to their rabbits from time to time. But many rabbit owners worry about offering the wrong treat and are fearful of giving their pet bunny something that could be potentially harmful to its health.

The main thing to remember when choosing any type of treat is that moderation is the key. Any type of treat that is given in excessive amounts could have the potential to be harmful to your rabbit because an abrupt change in diet can upset your rabbit's digestion. Therefore, always take care to offer treats only in small amounts so that you don't accidentally harm your rabbit.

There are many treats that are safe to feed your rabbit in small amounts, but some of the more popular choices are as follows: Carrots, apples (not the seeds), bread, dandelions, oatmeal, raspberries, strawberries, and bananas.

It's best to avoid any treats that are high in sugar content. Yogurt chips or drops are often sold as rabbit treats, but these are actually quite high in sugar and are not an ideal choice for rabbits.

Your house rabbit may quite possibly live in peace with your other household pets, but you'll need to be very watchful of your pets as they get to know one another.

RABBITS AND THEIR FRIENDS

Rabbits can (and do) coexist peacefully in a home situation with other pets. Rabbits and cats are often particularly well-suited as companions. A dog can also live in perfect harmony with a rabbit, although there is a bit more potential for a problematic situation unless you are perfectly sure of the dog's disposition. Rabbits often enjoy the company of another rabbit, although there can be exceptions, particularly depending upon the gender of your rabbits. It is unlikely that two males will get along with one another unless they are neutered. Two females often live together quite nicely, but they can annoy one another and fighting can occur, so watch carefully to make sure that they are co-existing peacefully. A male and female rabbit usually get along very well, however if the pair are your pets rather than a breeding pair, one or both will need to be altered in order to prevent the inevitable litters that will occur.

The most important thing is to be sure that you carefully observe the situation when you introduce your pets to each other. Be prepared for unexpected behavior, and be prepared to promptly step in if either (or both) animals show any sign of anger or unpleasantness toward one another. The last thing you want is for your pets to injure one another in their quest for dominance, so don't leave them alone unless you're positively sure that they are getting along and have become friends.

LITTER TRAINING

This is an area that you probably won't be exploring if you're raising rabbits in any quantity. Litter training simply isn't practical for large-scale rabbitries, but it is a very good idea for anyone with a couple of pet rabbits or for rabbits that are allowed to roam throughout the house.

Similar to litter-training a cat or housebreaking a dog, training your rabbit to use the litter box isn't

Litter training is fairly easy to accomplish with rabbits, but it does take time and patience. You will want to make sure that your rabbit is well trained to a litter box before you begin allowing the rabbit to play indoors.

Your rabbit's litter box will need to be lined with either wood shavings or pet litter. As with any pet's litter box, your rabbit's box will need to be regularly cleaned.

terribly difficult, but it does take time and patience. The end results are well worth the effort.

A simple way to begin the process of litter training is to observe your rabbit's current habits. Let's assume that you are keeping your rabbit in a wire hutch with a pan underneath. However, you would like to allow it to play in your living room, which you have carefully rabbit-proofed. Before allowing him free rein in your living room, take a careful look at the pan underneath your rabbit's cage. In all likelihood, you will notice that the droppings will be confined to a particular corner, or possibly two corners, of the pan. This is how you will determine where to initially place your rabbit's new litter box in its cage. If your rabbit is already used to using a particular location for defecating, then placing the litter box in this location is always wise. However, it's possible that your rabbit may avoid the new litter box and may choose a different corner. In this case, you may want to try to help your rabbit to get used to its new litter box by placing a few of its loose droppings directly into the litter box. This will help to convince the rabbit that the litter box has a familiar scent and may encourage it to begin using it. If this tactic is not successful, then you may want to try moving the litter box to a different location and continue experimenting until you have successful results.

Once your rabbit has begun using the litter box regularly, then it's possible for you to move the litter box to a new location and still have your rabbit understand that it is the location to use. In this manner, you may be able to move the litter box into your living room and have your rabbit use it. This works in your favor because the litter box already smells familiar to your rabbit and your living room does not. Therefore, the litter box is a much more appealing choice to most rabbits. Many people keep a couple of litter boxes available on different ends of a room to give the rabbit ample opportunity to utilize them rather than using the floor.

SAFE HANDLING OF RABBITS

When handling your pet rabbit, you have two main priorities: safety for your rabbit and safety for yourself. Obviously with respect to your own safety, we're referring to the potential of being scratched, which is a very valid possibility when handling rabbits. They can make a sudden move when you least expect it, and their super-sharp nails can easily come into contact with your skin. You want to minimize your risk of being scratched, which can be accomplished through careful handling. You also want to be very careful not to drop your rabbit, in order to protect him from possible injury. It's very easy to get flustered when a rabbit begins jumping or struggling to get away from you, particularly if you're being scratched in the process.

Many people lift their rabbits by the scruff of the neck, then carefully bring the rabbit toward their own body and support the rabbit in the crook of their arm against their chest. However, many rabbit breeders—particularly fancy rabbit enthusiasts—feel that you should never lift a rabbit by the scruff of its neck, as this can cause the flesh to loosen around the nape of the rabbit's neck and damage the rabbit's fur. Alternatively, you can lift a rabbit by leaning over it, carefully pulling it toward your body, and supporting it promptly with your arm around its body and underneath the rabbit.

In either case, the priority is to immediately support the rabbit so that it feels safe and secure. A rabbit that is fearful of being dropped is going to jump and flail about with its claws flying. It is not only an unpleasant and traumatic experience for the rabbit, it's also somewhat unpleasant and traumatic for you, particularly if your arms are getting scratched to pieces in the process. Always carry your rabbits by supporting them underneath and around their body to give them the comforted feeling of being safe and protected.

Toddlers love rabbits, but this is another situation that you will want to carefully monitor. Any toddler visiting your home should not be left unattended with your house rabbits, for the safety of both rabbit and child.

Always firmly support your rabbit by placing your hand or arm underneath him. Rabbits are fearful of falling and need to feel that they are being supported.

CHAPTER 8

MANAGING YOUR RABBITRY

The outward success of any rabbitry is fueled by a great deal of work behind the scenes. In this chapter, we'll discuss the importance of several areas that will help to make your rabbitry a success, from record keeping and marketing to evaluating your breeding program and choosing stock to upgrade. Raising rabbits is a continually evolving process that is influenced greatly by these factors and the choices you make. Let's take a closer look at the different areas in which you will manage your rabbitry.

Every breeder's goal is a top-quality rabbit exhibiting wonderful breed type. If the rabbits you are producing don't meet your ideal expectations, you may want to consider upgrading some of your breeding stock.

DOE BREEDING RECORD

NAME: _Cuddles_ TATTOO: _C-R109_

DOB: _4-26-06_

DATE BRED: _1-3-08_

DATE DUE: _2-2-08_

PREGNANCY CHECK: _OK_ ✓

DATE KINDLED: _2-1-08_

NUMBER OF LIVE KITS: _8_

SIRE: _Frolic C-R963_

This is a very simple and easily maintained type of breeding record. This sheet keeps track of the pertinent information for each doe in your rabbitry. With each successive litter from each doe, you will create a new sheet to record her information. Always save the old sheets and place them with your permanent records.

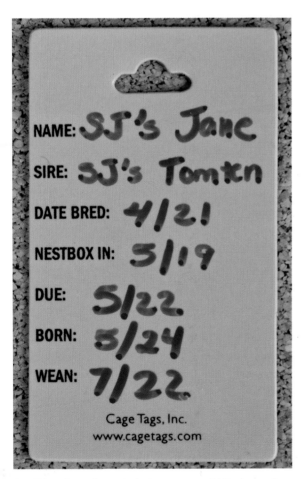

NAME: SJ's Jane

SIRE: SJ's Tomten

DATE BRED: 4/21

NESTBOX IN: 5/19

DUE: 5/22

BORN: 5/24

WEAN: 7/22

Cage Tags, Inc.
www.cagetags.com

Small hutch cards can help keep your rabbitry records in order and up-to-date. This small plastic card has space to record the doe's name, the buck she is bred to, breeding date, due date, delivery date, weaning date, and the date that the nest box should be added.

KEEPING RECORDS

You've just read the words "keeping records" and you're probably thinking: *Records . . . maybe I can skip over this part. Maintaining extensive records is probably only for those big, commercial rabbitries with hundreds of rabbits. Since I have only a dozen rabbits, I'll be able to remember everything without spending all of that time on record keeping. Right?*

Not quite. Record keeping, regardless of the size of your rabbitry, is truly vital to the mechanics of any rabbitry. With good records, you will be able to calculate kindling dates, track the production of your does, evaluate the production of your bucks, track pedigrees, investigate family lines to determine which ones are producing the best for your rabbitry, etc. The importance of good record keeping cannot be underestimated.

Traditionally, records have been kept by hand with forms painstakingly filled out in pencil or ink and stored in a binder or folder for easy reference. Even today, many breeders prefer this method of record keeping. Others prefer keeping their records on a computer. There are a multitude of computer software programs available that allow to you to enter all of your rabbits into the system, create pedigrees, record vital information such as registration numbers and tattoo numbers, and print reports. Some rabbitry owners find that this type of software allows them to keep records that are more accurate and up-to-date with a smaller time investment.

Pedigrees contain much more than just a list of names. These fully completed pedigrees also list tattoo numbers, colors, weights, and Grand Champion numbers (where applicable) for all of the ancestors listed on the pedigree.

If you are the owner of only a few animals, record keeping is simpler. For instance, if you have only one buck, there is no question of which buck your does are bred to. On the other hand, if you have a larger rabbitry with 20 does and 4 bucks, things start to get a bit more complicated. This is where you find the increased need for detailed and accurate record keeping. Never rely on your memory with regard to rabbit records, whether your rabbitry is large or small. It isn't possible to remember everything, and your records are too important to risk forgetting something.

Some of your most important records will undoubtedly be your breeding records. Without recording the date that your doe was bred, you will have no way to determine when she is due to kindle, when to add the nesting box, and when she should be palpated for pregnancy diagnosis. The extent of breeding records can vary considerably. Some breeders like to keep track of the number of kits delivered, the number of kits that survive, and the weight of the kits at three weeks of age. Others like

to keep simpler records that show only the breeding date, the date that the nest box should be added, and the date of delivery. Some breeders like to keep the breeding records directly on each doe's cage, as they can then tell at a glance which does need palpating for pregnancy diagnosis, which ones need their nest boxes added, which ones can be re-bred, etc.

Other records could include registration numbers, tattoos, medical records, financial records, or production records (how many litters produced by each doe and how many sired by each buck). Over time, you will be able to fine-tune your record-keeping system into the method that works best for you, regardless of whether you decide to invest in a computer program or manage your records by hand.

PEDIGREES

One of the main components of record keeping deserves more extensive coverage, and that is the subject of pedigrees. If you are breeding rabbits and want to achieve registered stock, you will be making yourself very familiar with pedigrees because you

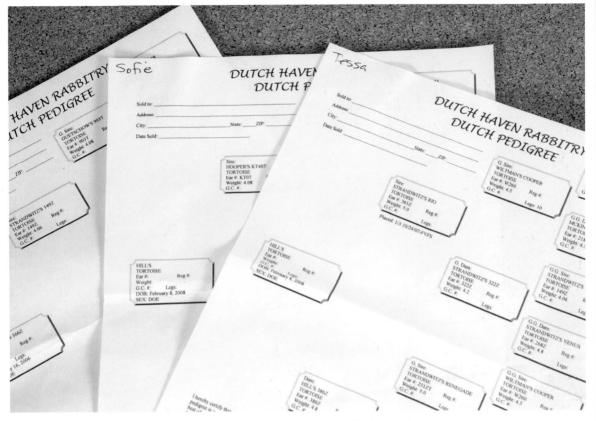

Computerized pedigrees are very easy to prepare and look quite professional. There are many computer programs specifically designed to help rabbitry owners record pedigrees and other information.

cannot register rabbits without complete pedigrees. Some people find pedigrees and registration to be very confusing subjects, and they certainly can be when you're first trying to understand them. Let's take a closer look at pedigrees and the process of registration.

A purebred rabbit is a rabbit whose ancestors for three generations are all composed of the same breed. This is as opposed to a mixed-breed rabbit, whose ancestors are composed of two or more breeds. Occasionally, infusions of outside breeds are sometimes used to improve or fine-tune a breed of rabbit, particularly in the developmental stages of new breeds or varieties. For instance, in the case of the American breed of rabbit, there are breeders undertaking the task of establishing the Red variety in addition to the Blue and White varieties already recognized by the ARBA. To achieve the ideal Red coloring, some breeders have used Thrianta rabbits crossed with their Americans in order to gain the Red coloring in the Americans. Obviously, the offspring cannot be considered to be purebred

for several generations to come because the entire three-generation pedigree must consist of only Americans. In the meantime, the rabbits may certainly be shown at ARBA shows, despite the fact that their pedigrees are not completely purebred. However, they cannot be registered, because a rabbit must have a purebred pedigree for three generations before the ARBA will register it. The three generations of animals need not be registered with the ARBA, which is an important concept to understand. The animals in your rabbit's pedigree simply need to be representatives of the breed, not necessarily registered.

In addition to the names of ancestors for three generations, a rabbit's pedigree will also include other vital information, such as color, weight, and ear tattoo numbers. If any of the animals have been Grand Champions, this is also noted on the pedigree. The ARBA issues different colored seals on the Certificate of Registration to signify how many rabbits in the pedigree are registered with the ARBA.

In order to be registered with the ARBA, a rabbit must meet several qualifications and be examined by a licensed ARBA registrar. One of the main requirements is that a rabbit must meet the standard weight for his or her breed, which is why all rabbits are weighed before they are allowed to apply for registration.

In order to be registered with the ARBA, your rabbit must be examined by an official ARBA registrar. Your rabbit will be evaluated against its breed's entry in the *Standard of Perfection*, and if it is found to be of sufficient quality and to meet all requirements without any disqualifications, then the registrar will fill out an application for your rabbit's registration with the ARBA. Rabbits must be over six months of age before they can be registered.

Pedigrees are also very valuable for purposes aside from registration. In order to make the best decisions for your breeding program, you will need to know which rabbits come from which bloodlines. For instance, let's say you have been raising rabbits for five years and have produced a two-year-old doe, Thumper, who has produced three large litters of excellent quality. Thumper is also a Best of Breed winner on the show table and is a top-quality doe in her own right. By checking your records and looking at Thumper's pedigree, you will be able to make decisions that may help you select for certain positive characteristics in future breedings.

You may decide to keep back three of Thumper's sisters for your herd in order to focus on those bloodlines in hopes of improving the overall quality of your litters. You may decide to try linebreeding by breeding Thumper or one of her sisters to their sire's full brother in an attempt to firmly establish the good characteristics of their bloodlines in your herd. Or maybe you'll decide to sell three or four of your does that are unrelated to Thumper in order to focus on rabbits with similar pedigrees and type.

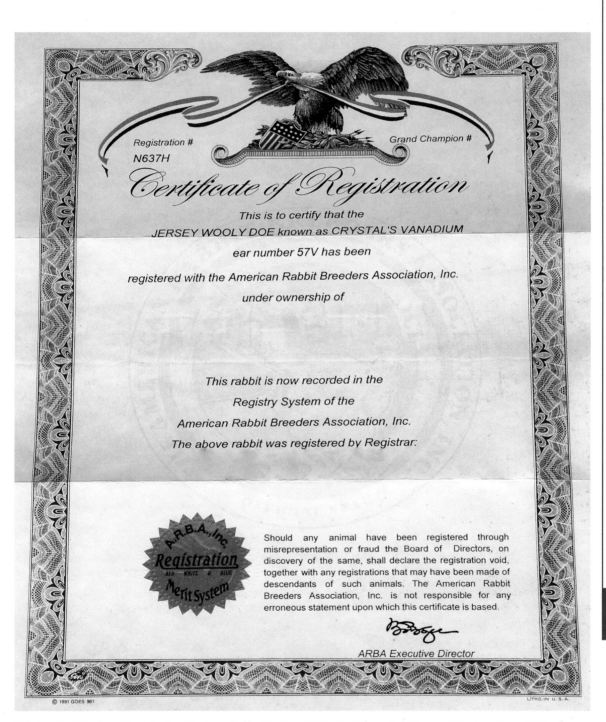

Registration #
N637H

Grand Champion #

Certificate of Registration

This is to certify that the

JERSEY WOOLY DOE known as CRYSTAL'S VANADIUM

ear number 57V has been

registered with the American Rabbit Breeders Association, Inc.

under ownership of

This rabbit is now recorded in the

Registry System of the

American Rabbit Breeders Association, Inc.

The above rabbit was registered by Registrar:

A.R.B.A., Inc.
Registration
RED WHITE & BLUE
Merit System

Should any animal have been registered through misrepresentation or fraud the Board of Directors, on discovery of the same, shall declare the registration void, together with any registrations that may have been made of descendants of such animals. The American Rabbit Breeders Association, Inc. is not responsible for any erroneous statement upon which this certificate is based.

ARBA Executive Director

© 1991 GOES 961

LITHO.IN U.S.A.

This is a Certificate of Registration that was issued by the ARBA. On the back of the certificate is the rabbit's pedigree, as well as information on the colors and weights of each of the rabbit's ancestors.

Pedigrees can also help you to select against certain undesirable characteristics that you will want to eliminate from your breeding program. If you discover that you are consistently producing rabbits with a particular fault or weakness, pedigrees can help you pinpoint the bloodlines that may be connected with the problem. The same is true for whatever positive or negative characteristic you are trying to select for or against, including characteristics beyond the physical, such as temperament, maternal instincts, and large litters.

All of this considered, it's also important to know when your breeding program is in need of an outcross. Linebreeding and inbreeding (a more intensive form of linebreeding) undoubtedly have their merits in a rabbit breeding program, but it's also important to acknowledge when outcrossing (choosing an animal from unrelated bloodlines). Continuing with our discussion of the fictitious Thumper bloodlines, let's say that you've realized that most of your does are somehow related to

Thumper and your two current herd bucks are her sire and his brother. In this case, when looking for an outcross, you would most likely look for a new buck of unrelated bloodlines in order to take your program to the next step.

This is where pedigrees come into play again. When searching for this very important addition to your program, you will probably look at many choices before settling on one to purchase. By studying pedigrees, you can compare the bucks you are considering not only by their looks, but also by breeding. Pedigrees can grant insight into whether a buck comes from lines that are notably of a consistently nice quality, or whether he is merely a nice-looking fluke that may or may not produce as well as he appears himself.

Pedigrees can be recorded by hand, as they have been for generations. The ARBA sells pedigree booklets for three dollars; each book comes with 48 blank pedigree pages. As mentioned previously, there are also many computer software programs

A clamp-style tattoo set complete with its accessories is shown here. The new pen-style tattoo kit is gaining interest among breeders, but many rabbitry owners still use the traditional clamp-style tattoos for their rabbits.

that allow you to easily input pedigrees into your computer and print them out. This results in a very professional-looking pedigree that is easily read (no more trying to decipher your own handwriting on a written sheet.). There are also websites that offer free pedigree services, such as www.sitstay.com. Do some searching and experimenting to find what works best for you.

TATTOOING

Now that we've discussed the importance of record keeping and the importance of pedigrees, it's time to turn our attention to another form of record keeping: tattooing. For as long as people have been domesticating rabbits, breeders have been using different ways to keep track of them. Over time, these forms of identification have evolved from holes punched in the ear to hanging tags to today's permanent tattooing with ink. Tattooing is essential for many reasons, not the least of which is the fact that no rabbit can be shown at an ARBA-sanctioned

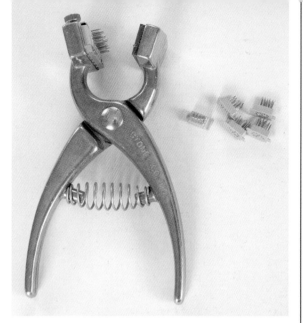

A tattoo kit usually comes with a set of letters or numbers, but additional sets of letters and numbers can be purchased if you wish to increase your options for tattooing.

Before you tattoo each rabbit, you will need to adjust the lettering or numbering on the tattoo in order to individualize each rabbit's tattoo.

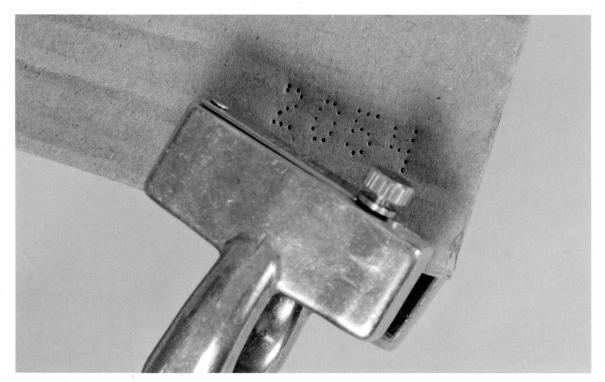

Once you've set up the lettering or numbering on your tattoo, it's always wise to run a test on a piece of cardboard in order to make sure that the tattoo design is in proper order. It's easy to put a letter or number in upside down by mistake, so double check before you officially tattoo your rabbit.

Before you actually tattoo your rabbit's ear, it's always a good idea to clean the inside of the rabbit's ear with a small alcohol wipe.

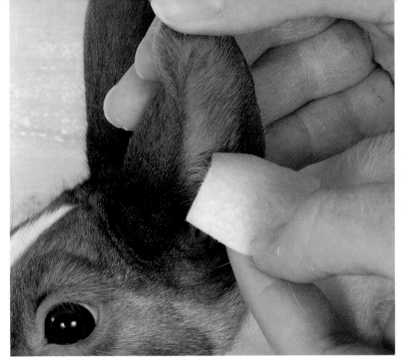

Thoroughly scrub the interior of the rabbit's ear and allow the area to dry before you begin tattooing.

Carefully restrain the rabbit and align the tattoo clamp along a portion of the rabbit's left ear.

show without a permanent tattoo in its left ear. (The right ear is reserved for tattooing the official ARBA registration number, in the event that the rabbit is registered. The left ear is the tattoo number/letter combination chosen by the breeder.) Tattooing is also important for identification purposes within your own rabbitry. Obviously, if you have only a few rabbits, you probably won't need to rely on tattoo identification in order to determine which rabbit is which. But you might be surprised at how quickly and easily a person can become confused, particularly if you have multiple rabbits of the same

Squeeze the clamp to complete the tattoo. After this, you will need to apply some ink to the area.

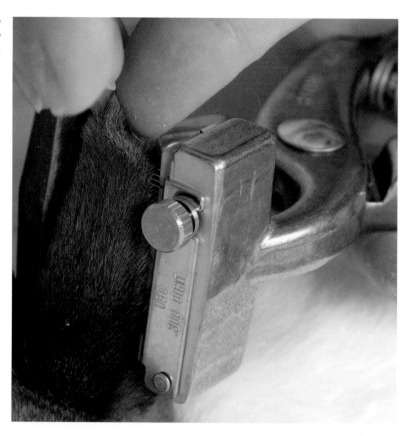

When correctly applied, the pen type of tattoo kit can produce legible results with less hassle than the traditional clamp-style tattoo kit. However, when incorrectly applied, the pen can produce irregular and smudged results, as seen here.

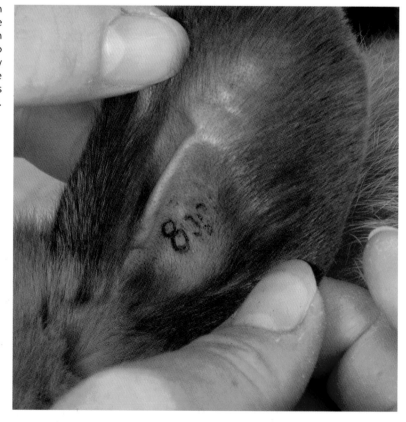

color pattern. Tattooing takes away any guesswork and gives you an accurate and permanent form of identification that will always tie the rabbit to its personal information, pedigree, etc.

You may wonder at how breeders establish their system of tattooing. What exactly is the significance of RC2102 or 14ARK? Every breeder's system is a little different. Many breeders use the initials of their rabbitry in their tattoo system. The Really Cute Rabbitry might start all of their tattoos with "RC," which will permanently identify all of their rabbits as having been born in the Really Cute Rabbitry. The numbers following the RC in RC2102 might stand for the 21st rabbit tattooed in 2002, or the 2,102nd rabbit born in their rabbitry. There is no right or wrong way to set up a tattooing system. Choose one that is easy for you to decipher and works well for you.

There are a couple of different methods of applying the actual tattoo: clamp-style tattooing and hand-tattooing with a pen. The clamp-style is the type that has traditionally been used, and is still quite commonly seen today due to the fact that it is an inexpensive and quick way to tattoo rabbits. The newer form of using a tattoo pen also has its benefits, including less pain for the rabbit and easier application for smaller breeds of rabbits with small ears. Tattooing equipment is available from rabbit supply companies; prices range from about $30 to $50, depending on the type of equipment that you choose.

Here is a very nice example of a well-done tattoo. A tattoo in a rabbit's left ear is necessary in order to show at ARBA-sanctioned shows and is very important for your own record-keeping purposes.

MARKETING

Marketing is the area in which some rabbitries excel and others dismally fail. Quite simply, marketing can significantly influence your rabbitry's success. After all, developing a wonderful breeding program and producing top-quality rabbits does not automatically result in sales if no one knows that your rabbitry exists. You must develop a marketing strategy that will ensure that buyers seek out your rabbits and come to your rabbitry when they are looking for rabbits to purchase.

This is perhaps easier said than done. Marketing can be a time-consuming and challenging task. It takes time to create ads, set up a website, take photographs, answer e-mails, and work on promotion. With today's hectic lifestyles, carving out time to market your rabbits can be hard to do. But there are some ways to make the job easier and more efficient.

PRINT ADVERTISING

Print advertising is the traditional form of marketing and offers many opportunities to get the word out about your rabbitry. There are a number of options for placing advertisements about your rabbits, so we'll discuss some of the most popular choices.

Domestic Rabbits Magazine

Domestic Rabbits, the ARBA's official publication, reaches every member of the ARBA, making it a good place to advertise if you want your ad to have maximum distribution among rabbit enthusiasts.

Specialty Club Publications

These are particularly good marketing resources if you're looking to market breeding stock. Advertising in the publication of a specialty breed club ensures that your ad is seen by other

Writing an ad is an art of its own. In a publication filled with classifieds, how can you make sure that your ad stands out? Engaging text and an eye-catching title will help to attract the attention of readers.

140

Looking to advertise? Magazines for rabbit enthusiasts are excellent places to tell the world about your rabbitry. Ad rates will vary depending on the type of publication, but most offer classified advertising that is an economical way to spread the word about your rabbits.

enthusiasts of the same breed who might be more likely to be interested in what you are selling. Advertising in a specialty club publication also helps to establish you as a serious breeder.

Local or Regional Club Publications

Also an exceptionally smart choice, advertising in the publications of your local or regional clubs allows your ad to be seen by other rabbit enthusiasts in your area, which can be very beneficial.

Newspapers

Don't overlook the obvious. Even though it isn't a rabbit-specific publication, your local newspaper can be a wonderful place to advertise and market your rabbits. There are often individuals looking for rabbits—sometimes for their children as pets or 4-H projects, or sometimes to begin raising rabbits. If you make it known that you have rabbits available, you might be very surprised by the local interest.

In any of these cases, it's also good to remember that marketing through print advertising is

something of a long-term commitment. A one-time ad in a newsletter may not produce much interest. To be truly effective, your print advertisements will need to appear regularly in your chosen advertising venue, probably over the course of several months or a year. This can be a considerable financial expenditure, positive increases in sales should follow. Pay close attention to which ads garner the most interest and shape your future advertising plans to include the publications that result in the most sales.

Internet Advertising

The Internet has literally changed the face of marketing in the rabbit world. E-mail, websites, blogs, and discussion forums have become major forms of networking and advertising. People can advertise rabbits on the Internet, post photographs and descriptions, answer e-mails, and may even have sales by the end of the day. For many people, speed and convenience are important priorities, and the Internet provides both.

A website for your rabbitry can be a great way to attract new customers, especially if you keep it up-to-date and have good photography depicting your rabbits.

So how do you capitalize on the opportunities that the Internet offers for rabbit owners? By letting people know that you're out there. This can be accomplished in many ways, but probably one of the easiest is to join a rabbit discussion forum, such as the ones offered through Yahoo Groups. Choose a group that closely fits your interests, such as a group for enthusiasts of your breed or a group that connects people from your area; then join in on the discussions. Many forums allow you to post pictures of your rabbits, share your plans for your breeding program, or list rabbits for sale. This is a good way to start getting the word out about your rabbitry. Make sure that you promptly respond to any inquiries about your rabbits, whether it's a person commenting on your cute photo or someone interested in a rabbit that you have for sale. Promptness and pleasantness go a long way toward establishing a positive reputation.

The Internet and e-mail are efficient ways to keep in touch with clients and to get the word out about developments in your breeding program. You can quickly alert potential buyers when you have a litter that's ready to go or you've decided to offer some senior rabbits for sale.

Of course, hand-in-hand with the Internet and e-mail are websites. You may find that a website is one of the most valuable marketing tools that you can utilize. A website can be as simple or as complex as you choose, but there are some key points that will help to make your website user-friendly and a positive addition to your marketing program.

- **A picture is worth a thousand words**
  One of the most important attributes of your website will be the photographs. Although it will take time and effort, your website will benefit from attractive, high-quality photography that depicts your rabbits in a positive way. A portrait of each member of your breeding herd is a very good way to show potential customers the quality of your breeding program. Pictures are also vital when you have a rabbit (or rabbits) for sale. People like to see what they are buying, so photographs are a necessity.

- **Be friendly**
  Chances are, the people who visit your website have either stumbled across it on a search engine or found it through a link from some

other source. In either case, your website is one of thousands to be found on the Internet, so it will need something to make it stand out from the others—something to make people interested enough to look around for a few minutes before clicking on to another site. One of the main ways to do this is to exude friendliness throughout your website. An "about me" page is always a plus. Describe your personal interest in rabbits, how long you've had them, how your breeding program was developed, ways that you're working on improving it, show successes, your favorite rabbits in your herd, etc.

- **Double-check**
Before you upload your site, and each time you make changes or updates to a page, take a few moments to reread what you've written. Is it as clear as it can be? Are there any typographical errors that need fixing? Take a minute to run the spell-checker to catch any misspelled words.

Typos and spelling errors can detract from the overall impression of your website.

- **Keep it updated**
Completing a website is quite an accomplishment, but it doesn't end there. You will want to make sure that your site stays up-to-date. Keep it fresh by updating information and news, posting announcements about new litters, updating with new photographs as necessary, and generally changing anything that looks old or outdated. Also, be sure to change the date on your "Site Last Updated" marker at the bottom of your homepage. A person visiting your website is much more likely to pass it over if the "Last Updated" date is months (or years) old. Sites that have been recently updated are much more likely to capture someone's attention. A recent update assures visitors that the information is current, and they won't have to worry about wasting their time by inquiring about rabbits that have long since been sold.

SETTING UP A WEBSITE

Many people are overwhelmed by the idea of creating their own website, but the process can really be quite simple. There are many types of computer software that allow you to easily create a simple website without having to learn all of the intricacies of the HTML programming language. On the other hand, if you're the techno-savvy type, you can create a more complex website using HTML code. Or you can hire a website designer to create a website for you. This will likely result in a very professional-looking site, but you must be prepared for the cost involved with hiring a designer. In addition to the design and setup charges, you may also be charged for maintenance each time you want or need the site updated. Handling your own website maintenance is one of the advantages to designing your website yourself.

You will also need to decide whether to have your website hosted by a free server or to purchase a registered domain name for your website. Obviously, the distinct difference between these two options is the cost; free is free, as opposed to a site that requires an initial registration charge followed by monthly fees. For many rabbit owners, their website needs are amply met by a free website. The downside to the free sites is that your website address can get quite lengthy. These long addresses can be harder for people to remember, which is a definite marketing disadvantage. Free sites often have limits on daily traffic, which can be a problem if your website attracts high traffic, as some potential viewers may be unable to view your site at certain times of the day. On the other hand, if you register your own domain name, you can have a web address that is shorter and easy to remember, such as www.mygreatrabbitry.com. You can also have unlimited visitors, and no one will get turned away from your site because of high traffic.

CLUBS AND SHOWS

Another way to increase awareness about your rabbits is through affiliations with clubs and participation at shows. As with promoting any product, networking is very important; you can make progress in this area by getting acquainted with other rabbit enthusiasts, exhibiting at shows, and becoming a member of your local breed clubs. These are excellent ways to increase awareness for your rabbitry and allow people to see what kind of stock you have available. As your rabbits become better known, owners can point buyers in your direction. Remember that networking works both ways. If you have a buyer who is looking for something that you don't have, never hesitate to point him or her in the direction of another rabbitry that does.

Don't underestimate the importance of word-of-mouth advertising. If people are talking positively and recommending your rabbits to others, it can go a long way toward establishing the success of your rabbitry. Word of mouth is one of the most effective and productive means of marketing and is always one of the most positive ways to increase business.

Exhibiting at shows can also be a wonderful method to establish and promote your rabbits. Show wins help validate the quality of your breeding program. On the other hand, if your rabbits are consistently placing low (or not placing at all) in competition, their performance can help you fine-tune and evaluate where your stock may need improvement.

IMPROVING YOUR BREEDING PROGRAM

One of the fascinating aspects of rabbit raising is that it is a process. No breeding program ever reaches the pinnacle of perfection in which the breeder feels that there is no room for improvement in the herd. There will always be areas in which you would like to see improvement, slight

Colorful signs can draw attention to your sales rabbits when you're marketing them at a show. Clearly describe what you have for sale and use bright colors and bold penmanship to catch your customer's attention.

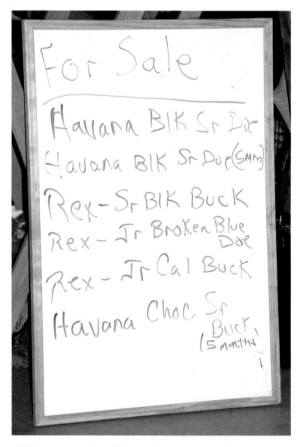

This is an example of a less-effective type of marketing. This sign, while factual and informational, does not have the pizzazz to attract the attention of people passing by.

144

Another excellent way to market your rabbits locally is at a small animal swap meet. For a slight fee you are allowed to set up for the day with your rabbits and market them to customers attending the event.

changes you would like to make, and qualities that you would like to focus on more thoroughly. In a nutshell, continued improvement is part of the driving force that keeps people enthused about rabbits for years (or decades).

Let's say that you've started with a trio of Mini Rex rabbits. On the whole, they are good-quality individuals and have placed well on the table, but they aren't quite the caliber to win any Best of Breed (BOB) or Best of Show (BOS) awards. Looking critically at your buck, you have to admit that his ears are not quite as balanced as they could be, but he has excellent shoulders and nice hindquarters. One of your does is also quite nice in the shoulders, but her hindquarters are a bit narrow. The other doe has narrow hindquarters but the benefit of an exceptionally gorgeous head.

Now let's say that you are evaluating the first litter from each doe in order to decide which ones to retain for your breeding program and which ones to sell or otherwise cull from your program. From your first doe, you have a litter of six that consists of two bucks and four does. One buck has unbalanced ears and narrow hindquarters

(a fault from each parent), so you slate him for sale as a pet-quality rabbit. The second buck has superb shoulders and nice hindquarters, plus very nice ears. You decide to keep him around for a while as a junior buck to watch him mature. Of the four does, two have narrow hindquarters and two have good-quality hindquarters, all have nice shoulders, and only one has unbalanced ears. You decide to keep the very nicest doe.

In the second litter there are four rabbits: three does that inherited their mother's beautiful head and one buck with a plainer head. The entire litter inherited their father's quality shoulders and hindquarters. Out of this litter, you are impressed enough to keep all three does. Thus, out of your first two litters, you have kept four does and one buck for your future program. This gives you a nice foundation herd of six does and two bucks, all of which are relatively good examples of the breed, yet each with areas where you could desire slight improvement.

As time goes by, additional litters are born. As with previous litters, there are some individuals that are excellent examples, others that are average, and

In order to continue improving your breeding program, it's important to take the time to evaluate the young rabbits that you're producing. Although this kit is only three weeks old, he is already exhibiting certain characteristics that are very appealing.

some that are substandard. By now you've probably been out to some shows, putting your stock up against others in pursuit of success, and had the opportunity to view rabbits that are winning on the show table. Perhaps you've even talked with some judges about the ideal qualities of the Mini Rex breed and have carefully considered these points.

All of this brings us to the next stage of your program: the improvement stage. This is the point where you stand back and critically evaluate your rabbits and the offspring they are producing in an attempt to determine where and how your program might be improved.

This stage is an important crossroads because while you want to continue improving in areas where your rabbits might be lacking, you also want to be careful not to backtrack from any progress that you've achieved. For instance, if you've carefully worked on selecting for good shoulders, you don't want to sacrifice this progress by choosing a new buck that lacks in shoulder quality but has perfect ears (the next area that you want to improve). You

will want to keep looking until you find a buck with the ears you're seeking, along with the same good shoulders that you've worked so hard to select for in your herd.

Don't be afraid to pay a bit extra when you find a rabbit that is perfect for the next step of improving your program. The added investment, when spread across a few years and many litters, will be a negligible expense if it means that your overall program is heading in a positive direction.

Upgrading your stock is a continual process as you fine-tune and focus in on your ideal characteristics. As time goes by, it's also possible that you may discover certain undesirable hereditary conditions that may crop up in your herd on occasion, such as malocclusion of the teeth and improperly colored toenails. If the problem is extensive and shows up often in your litters, you may want to carefully evaluate the pedigrees of your stock to determine the source in case you decide to isolate the carriers and potentially remove them from your herd.

If you are working with a breed that has very distinct color or color pattern requirements, you may discover that some of your stock consistently produces undesirable colors or patterns. In this case, you may want to concentrate on the pedigree lines that are more regularly producing the ideal color characteristics that you are seeking. A good working knowledge of rabbit color genetics will help you in isolating improper colors if you can determine which rabbits are carrying the undesirable colors.

Next are the production reports. As we've already discussed, solid record keeping is essential in order to evaluate the production of your rabbitry. For instance, suppose that when comparing the production records of two does, you realize that one produced four litters in a year and the other produced only two. It only makes sense that the doe that produced more litters is a more valuable member of your breeding herd. Similarly, if a doe consistently produces litters with few kits, or if only one or two of her kits survive to maturity, that is another consideration when evaluating your does. Maternal instincts are also an area to consider, including seemingly trivial qualities like birthing in the nest box. These become very important characteristics if you've ever lost a litter due to the doe leaving the kits on the wire. In the case of a doe that has proven herself to be a good producer of multiple kits and a naturally caring mother, you may want to consider keeping her daughters for your program and perhaps culling any of your does that are not as maternally inclined.

Production reports for your bucks should also be evaluated. If one isn't producing as you feel he should be, then you may want to consider replacing him. Of course, production has to be considered along with all of the other characteristics that we've previously discussed, including breed type, quality, and coloring. No decision to cull should ever be made without thoroughly evaluating all aspects of the animal's impact on your breeding program.

At first glance, these young tortoise Dutch does appear to be mirror images of one another. But a well-informed breeder can immediately spot the obvious differences and distinctions between the two rabbits.

CHAPTER 9
KINDLING, KITS, AND CARE

Now you are ready. You've researched the different breeds, invested in the best-quality breeding stock available, built a rabbitry, and set up all of the equipment. You've researched the various types of feed options, you have a good handle on what it takes to keep rabbits healthy, and now you are ready to begin pursuing your dream of raising rabbits and producing your first litters. However, even with all of your study and research, it's understandable that you may be a little nervous about entering the new areas of breeding and kindling. Don't worry, you will do just fine. Let's explore some of the basics of kindling, kits, and care.

This is the picture for which all rabbit breeders strive. A delightful litter of charming kits are adorable with their tiny whiskers and large eyes.

When breeding, it's always essential that you take the doe to the buck's cage and never the other way around. When the breeding is complete, you will want to remove the doe promptly and return her to her own cage.

This doe is 14 days pregnant, but you would never be able to tell simply by looking at her. Unlike many animals, a rabbit does not change very much in appearance when she is expecting.

BREEDING

Obviously, you will begin with a buck and a doe. Unlike many other species, there is some confusion regarding the exact science that dictates a doe's reproductive cycle. Some sources claim that a female rabbit has a heat cycle just as many other species do and that there is only a specific period of time during each cycle in which she is fertile. Other sources reiterate the traditional understanding that female rabbits are "induced ovulators," which means that they ovulate upon breeding. No one has definitively proven the answer to this age-old debate, but suffice to say that in either case, rabbits are incredibly fertile creatures, with the majority of all breedings (approximately 85 percent) resulting in pregnancy.

As your doe nears the end of her pregnancy, it's always a good idea to minimize her stress levels and to keep things quiet around the rabbitry. This might include keeping noisy children away from her cage if they seem to be bothering her.

For the actual breeding, you should always bring the doe to the buck's cage. This is one area in which all rabbit breeders are in agreement. You never want to bring a buck into a doe's cage because does are very territorial, and she could cause the buck significant harm if she views him as an infiltrator into her domain. Always remember to bring the doe to the buck and never vice versa. The actual breeding process only takes a few moments, after which time you can remove the doe and return her to her own cage.

You may want to consider breeding the pair again an hour or so later, and then again later in the afternoon (approximately 8 to 10 hours after the initial breeding). Many breeders feel that this significantly increases the odds of pregnancy, so even if you decline to do the second breeding in the morning, you will definitely want to do the afternoon breeding. Some breeders choose to leave the buck and doe together for an hour or so, while others claim to leave the pair together for several days. However, the majority of breeders leave the buck and doe together for only a short time and then separate them.

GESTATION

Many of the mammalian species, such as humans and horses, are very unpredictable as to their length of gestation. Who hasn't heard stories of women who have gone three weeks overdue before delivering their baby, or horse owners who are forced to endure a month of sleepless nights on foal watch while their pregnant mares simply stand and eat? Rabbits, on the other hand, deliver their kits after a gestation length of remarkable precision. Nearly all female rabbits kindle between 28 and 34 days, with the vast majority kindling on the 30th or 31st day. This makes it very easy to predict when the blessed event should occur and to eliminate any guesswork as to when the nest box should be added to the doe's cage.

Because you can predict a kindling date with a certain amount of accuracy, it's possible to coordinate the arrival of your litters at the most convenient time in your schedule. For instance, if you're going away for a weekend on the 15th and 16th of April, you wouldn't want to be breeding your does anywhere near the 16th to 18th of March, due to their projected kindling dates. By coordinating your does' breeding dates in order for them to deliver at the dates most convenient for you, you can help to increase the odds of safety and well-being of your kits. You will be on hand to assist if there is any problem, such as a delivery on the wire instead of in the nest box.

Some rabbit breeds are noted for having somewhat longer gestations than others. And, of course, not all does have "read the book." It's not at all uncommon for kits to be born on the 32nd or 33rd day, particularly with maiden does (first-time moms).

As a doe's pregnancy advances, you might notice a distinct increase in her appetite. This is completely normal, and you can certainly increase her rations until she delivers her kits. At that point, you will want to reduce her feed for a couple of days as she begins nursing her litter.

TWO PINK LINES . . . FOR RABBITS?

Your doe has been bred, and 14 days have ticked by on the calendar. You're wondering: Is she pregnant or isn't she? Since you can't run down to the drugstore to pick up a home pregnancy test for a bunny, you're going to have to rely on other, more primitive ways to ascertain your doe's pregnancy status.

Traditionally, breeders would attempt to diagnose pregnancy by trying a test mating, which means that they would test the doe to see if she was receptive for mating at approximately two weeks past the initial breeding. The belief was that if she showed interest in the buck, it indicated that she was not pregnant. If she wanted nothing to do with him, the doe was deemed pregnant. This method certainly works in some cases, but it is not as reliable as most breeders would like. Open does (those that are not pregnant) may not show interest in the buck simply because they are not interested, while pregnant does may be receptive despite their pregnancy. The latter can result in the possibility of a double breeding, which breeders would prefer to avoid.

Weight gain is another indication of pregnancy, but it lacks reliability. Weight gain cannot always be attributed to pregnancy, although in many cases it can be. It is an indicator that should be considered along with other factors, but lacks reliability on its own. If you decide that you would like to use weight gain as a possible pregnancy predictor, then you will need to carefully weigh your doe prior to breeding and maintain her daily rations without change for at least two weeks. If, at this point, you discover that she has gained weight, it is certainly an indication that she may be pregnant.

Palpation is considered to be the best and most accurate form of pregnancy diagnosis for rabbits. Around the 14th day of gestation, the doe's abdomen can be carefully palpated externally with your fingers. If she is pregnant, then the developing kits can be felt along either side of her abdomen. At this stage, the kits feel like hard marbles or grapes and are found in two parallel lines extending up the doe's body. If the palpation results in the assurance of pregnancy, then congratulations are in order.

If the palpation is negative, then most breeders will immediately rebreed the doe for another try. Always remember to record the date of this second breeding so that you can palpate again in 14 days and add the nest box on the 27th day after the second breeding. Obviously, you should never totally disregard the first breeding date, even if the palpation result was negative; keep it in mind so that you don't miss an unexpected litter around the 31st day from the first breeding.

Learning to properly palpate a doe can take a little bit of practice in order to master the technique, so if you can observe a knowledgeable breeder as he or she demonstrates the proper way to palpate, you can get a better understanding of exactly how to (and how not to) do the job. It is very valuable knowledge that will serve you well in your rabbitry for years to come.

Careful observation of your doe's behavior can also produce indications of pregnancy. If you routinely feed hay to your rabbits as a part of their daily diet, you may notice that your pregnant does will begin carrying around mouthfuls of hay as their pregnancies reach the 22nd to 27th day. This is a sort of "pre-nesting" behavior and is a fairly reliable indicator of pregnancy if you're still not certain.

TYPES OF NEST BOXES

Not all nest boxes are the same. You may want to experiment with a couple of different types until you settle on the variety that works best for your rabbitry. Wooden nest boxes have the advantage of being warm, which is always a plus in the winter months when you're trying to keep your litter as warm as possible. Always remember that rabbits love to chew on wood, so your wooden nest boxes may be the target of a bored doe's entertainment for an afternoon. Steel nest boxes are also widely used and are the type most commonly offered from rabbit equipment suppliers. Steel nest boxes are lightweight, easy to maneuver, easy to disinfect, and the bottom is often made of removable plywood or particle board for easy cleaning. One disadvantage to the steel nest boxes in my opinion is that the corners can be a bit sharp, so keep an eye out for that. The third type of nest box is made of wire, is rectangular shaped, and has an open top. These are lined with disposable liners that are discarded after each litter.

If you choose an aluminum nest box, it's wise to choose one that has a removable plywood bottom with holes. This allows for drainage and helps keep the nest box dry.

ABOVE: A few inches of wood shavings have been added to this nest box, and now it is ready to be filled with hay.

LEFT: Wood shavings—not sawdust—can be used to line a nest box. Sawdust is much too fine and dusty to be a safe choice for rabbit bedding.

Steel nest boxes are produced in a variety of sizes, so you can select the size that best suits your breed of rabbit. Generally speaking, it's wise to purchase a nest box that is only a few inches larger than your rabbit so that the nest will encompass the majority of the space in the box and leave the doe without room to soil the box. Wire boxes are usually found in one standard size of 18 × 10 × 8 inches, which is suitable for most breeds.

CARE AND FEEDING OF THE PREGNANT DOE

After pregnancy is confirmed, either by palpation or other means, you probably have about two weeks until the kits are delivered. What special care will your doe need in the meantime?

In the first place, if she is currently nursing a litter, this is a good time to wean them. If you wean the doe's previous litter when she is approximately two weeks away from delivering her next litter, she'll have time to rest a bit before her next litter arrives.

The bottoms of your nest boxes should be lined with an absorbent type of bedding. Wood shavings are a popular choice, but shredded newspaper is also used. Hay or straw should then be added on top of the bedding.

Now the hay has been added and the box is ready to be added to a pregnant doe's cage. Most does will rearrange the contents of the box until it suits her fancy, and then she will pull her fur to complete her nest.

I will occasionally provide a nest box with wood shavings and place the loose hay on the cage floor next to the box. Some does seem to prefer building their nest themselves and will spend a couple of days busily installing the hay into the nest box. This is fun to watch and gives the doe an enjoyable occupation.

In caring for your pregnant does, always remember the importance of peace and quiet. In order for your pregnant does to feel relaxed and safe, they need the benefits of a quiet atmosphere, one that is free of stressful conditions and anything that might cause them alarm. This means that you might consider keeping strangers away from the hutches and keeping loud noises to a minimum as your does near delivery.

Many breeders slightly increase the doe's daily rations as she nears her due date to compensate for the increased demands on the doe's body as her kits are growing rapidly in utero. However, at kindling time, the doe's pellet rations are usually decreased significantly, and she is kept on a reduced ration for the first couple of days post-kindling. This is to reduce the risk of conditions such as mastitis and caked breast. Once her litter is a few days old, you can begin to increase her ration of pellets so that she is built up to free-choice pellets by the time the litter is about 10 days old. Lactating does need more feed than usual to support their growing litters.

It is absolutely imperative that a pregnant doe have continual access to fresh, clean water. This is even more vital after her kits are born, as nursing will rapidly deplete a doe's hydration and she will need plenty of fresh water in order to avoid dehydrating herself.

One of the most important things that you will need to do for your pregnant doe is to provide her with a nest box on the 27th or 28th day of gestation. This nest box is often lined with a layer of wood shavings (not sawdust), then filled with hay or straw, although some breeders use shredded paper underneath the hay. The nest box provides a warm, dry, and safe place for the doe to deliver her young. The consensus among breeders is that the 27th day is typically the ideal date to place the nest box in the doe's cage, as it gives her a couple of days to begin preparing her nest and lining it with her own fur. This timing also prevents her from dirtying the box, which can occur if you place it in her cage prematurely. In addition, if your doe happens to deliver her litter a bit early, perhaps on the 29th day, then your nest box will be in place and prepared for the great event.

NEST BOX BLUES

It sounds so straightforward: on the 27th or 28th day of gestation, add the nest box to your doe's cage. Simple and easy, right? In most cases, the answer is yes. In some cases, it's not quite so easy.

You might be surprised at how many pregnant does view the nest box as a super-fancy and comfy-cozy litter box and promptly proceed to utilize the structure as such. This can be a definite problem, as you want the nest box to be a clean and healthy place for the kits to be born, not a messy box filled with urine and manure. If you discover that your doe is using the box for unsanitary purposes, you can try a couple of different options. The first step is to move the nest box to a different part of her cage on the off-chance that you inadvertently positioned the nest box over her natural litter area. If you move the nest box and find that she is still using it as a litter box, then you may want to try adding a second nest box to her cage. In this case, your doe will hopefully choose one box for litter and use the other for nesting purposes.

If you live in a very cold climate, you might like to provide a nest box for a good part of the year, and this might overlap with the time for a doe to deliver her kits. So, what to do if the nest box is already in the cage and you don't need to add it on the 27th day? Cleanliness is your number-one priority. As the 27th day of gestation approaches, make sure that the nest box is scrupulously clean. Replace any soiled bedding and continue to do so as often as necessary in order to keep the nesting area as clean as can be.

There are some cases when does (particularly maiden does) will absolutely refuse to have anything to do with the nest box. It simply is not their idea of a suitable birthing structure, and they may proceed (quite persistently) to build their nest on the wire floor of their cage. There are a couple of options in this situation. You can match her persistence and repeatedly move her nest-in-progress to the nest box, or you can provide her with ample nesting material (lots of hay or straw) throughout her cage so that she can choose her desired location and construct her nest there. Then you can either move the finished nest and the litter to the nest box after they're born or let her raise them in the nest on the cage floor. In the latter case, you will need to be absolutely certain that there is plenty of thick bedding along the wire floor to prevent any kits from getting a body part caught in the cracks or escaping altogether. Make sure that you line the lower portion of the cage walls for the same purpose. You can also line the walls with thin plywood or purchase special cage inserts from a rabbit equipment supplier. The important thing is to keep the kits in the nest. They cannot maintain their proper body temperatures if they are strewn around the cage or if they fall out of the cage and onto the floor of your rabbitry.

The nesting instinct is very powerful for does, and you want to work with their nesting plans whenever possible, rather than against them. Of course, there are wonderfully cooperative does who are delighted when you present them with their nest box and proceed to spend several happy days decorating their nursery with hay and fur. These does wouldn't dream of soiling their beautiful nest box with litter; they always keep it squeaky clean. Oh, to have a dozen of these marvelous mothers in our rabbitries.

KINDLING

Now we've come to the sneaky part: kindling time. Does are notoriously secretive when it comes to delivering their kits, and chances are good that kindling will occur during the night or early morning hours. Chances are even better that you will miss the entire event. It's important to respect the doe's need for peace and quiet during this crucial period. On the 29th, 30th, and 31st days of gestation, keep your intrusions to a minimum. By all means, check on your doe often during this period, but be quiet, cautious, and careful. If you find that she has delivered the kits on the wire floor, you will need to promptly deposit all of them into the nest box (as long as they are still alive), surround them with fur, and place them on straw. In the case that the kits are alive, yet are cold and chilled, it is possible to carefully warm them by taking them indoors and wrapping them in a warm towel. Be careful that you don't accidentally smother them. Some breeders like to wet the towels with warm water before warming the kits, and others will hold the kits directly under a faucet with the warm water turned on. If you choose this method, be careful that you don't inadvertently drown them. By warming the kits promptly, breeders are often able to save kits that would otherwise have frozen if left alone.

It's snowing lilac fur. Or more accurately, it's the day of delivery for this Rex doe. In addition to her nest, she filled her entire cage (and the floor surrounding her cage) with ample amounts of fur that she pulled from her body.

Just a few hours after birth, this litter of Rex kits is resting comfortably in their warm nest. The nest is composed of hay and fur.

All disasters aside, your doe will probably deliver inside her nest box as expected. Once you discover that she has delivered her kits, you'll want to subtly take a quick count in the nest box to see how many kits were delivered and make sure that all are alive and warm. In the unfortunate event that one or more of the kits are stillborn, you will want to remove them from the nest. You will want to peek in the nest box at least once a day to make sure that the kits are being fed (look for those full bellies) and that all are still alive.

Dystocia (malpresentation in delivery) is uncommon in rabbits, but nonetheless you will want to diligently watch your does as they near their time of delivery. If a doe has a difficult birth, she may scatter the kits throughout her hutch, due to being distracted and upset by the problem with delivery.

At one day of age, this kit is quite helpless. Kits are born blind, deaf, and hairless, and need the warmth that comes from the nest and the rest of the litter.

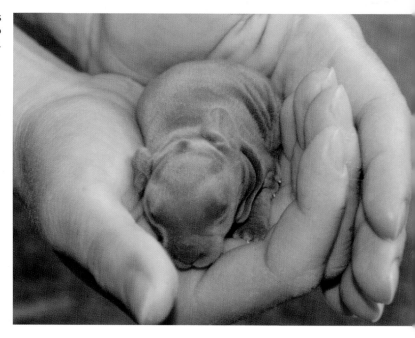

By three days of age, the kit has grown a bit and is also starting to get the slightest bit of fur.

At 10 days of age, this kit's eyes have begun to open, although he still holds them shut for the majority of the time. In addition, his ears have perked up and he's getting quite a bit more rabbit-like in appearance.

At two weeks, this kit is getting more and more curious and is ready to begin exploring the world.

Cannibalism is another problem that is occasionally seen, so it bears mentioning even though it is far from a pleasant topic of discussion. From time to time, you may find that a doe has overexuberantly welcomed her kits into the world, and you may discover kits that are missing portions of their body. Ears are commonly targeted in these instances, and it's not entirely uncommon to find a kit that is missing an ear (or two) or a portion of an ear. Sometimes the kits are damaged to the point that they do not survive. It is not known why some does display this type of behavior, but it is a valid reason for culling a doe, particularly if she repeats the tendency with more than one litter. Some breeders will excuse a solitary case of cannibalism, particularly if extenuating circumstances (loud noises in the rabbitry, something that may have frightened the doe, a first-time mother, etc.) may

have contributed to the problem. In these cases, the breeders will often give the doe a second chance with another litter, but if she repeats this unpleasant behavior, it is usually time for the doe to be culled from your herd. There is no need to perpetuate this kind of attitude and negative mothering ability in your rabbitry.

FOSTERING KITS

There is a wide variety in the number of kits that are born in a litter. Sometimes there are as many as 10 or 12 (the record is 22 kits), and sometimes there are only one or two. Some of this variation depends upon the breed of rabbit (generally speaking, larger breeds produce higher numbers of kits than smaller breeds), but it can also depend upon the individual rabbit and of course the individual litter.

The process of fostering kits occurs when a breeder has multiple does due to kindle at the same time. In the event that one of the does produces an exceptionally large litter—perhaps over eight kits—the breeder may remove a couple of the newborn kits from the nest box and transfer them to the nest of another recently kindled doe who produced a litter with fewer kits. Most does will not notice the sudden intruders and will nurse them along with their own young. This allows the doe with the larger litter to devote more of her resources (i.e., milk) to the remaining kits, while the fostered kits benefit from a more ample milk supply from the doe with fewer kits in her litter.

If a doe produces a very small litter of only one or two kits, then some breeders will foster these kits into another litter and effectively combine two small litters into one larger one. This is mainly for purposes of warmth for the kits, particularly if the litters were born in the winter. The body heat from multiple kits sharing a nest does wonders for keeping the entire group warm (and alive). A single kit or even two kits may have a hard time generating enough warmth to survive.

By three weeks of age, this kit spends less of the time sleeping and much more time playing, jumping, and running. The transformation from the first day is quite amazing to observe.

Fostering can also occur in the unfortunate event that a doe dies shortly after delivering her kits. The orphaned kits can then be distributed among a couple of other does that are nursing their own litters. For this reason, many rabbit breeders prefer to always have multiple does due to kindle at the same time, allowing for ample opportunities for fostering should the need arise.

One important note is to always keep records on which kits have been fostered. Without accurate knowledge of which kits came from each litter, you cannot evaluate which cross has produced the best kits, which is a necessary step in the continual improvement of your breeding program. Additionally, you cannot accurately list the pedigrees if you aren't positive of the parentage of your kits, so keep good records and mark the fostered kits (you can use a marker in their ear) if necessary to prevent confusion.

"PEANUT" KITS

This is a subject to be aware of if you are breeding any of the dwarf breeds, which include the Netherland Dwarf, the Holland Lop, the Britannia Petite, the Dwarf Hotot, the Jersey Wooly, and the Polish. These breeds carry the dwarf gene, which is inherited through simple dominance. This means that because the dwarf gene is a dominant gene, when it is carried in the genotype (genetic constitution), it is expressed in the phenotype (physical traits). Because genes are always found in pairs (called alleles), there are three possible combinations of the dwarf gene: 1. Dwarf/non-dwarf (known as a true dwarf), 2. Non-dwarf/non-dwarf (known as a false dwarf), and 3. dwarf/dwarf (known as a peanut, due to its miniscule size). Generally speaking, the majority of dwarf rabbits that you will see on the show table are dwarf/non-dwarf, which means that they exhibit the appearance of a dwarf but also carry the non-dwarf gene. This is the genetic combination that breeders of dwarf rabbits strive for.

Let's say that you are crossing two of your Dwarf Hotots, each of which are genetically true dwarfs. From this mating, you are presented with a litter of four kits. Theoretically, in this litter you will

Netherland Dwarfs, like the other dwarf breeds, can produce what are known as peanut kits. This is a lethal condition that occurs when a kit possesses two copies of the dwarfing gene.

A doe and her kits need more attention than some of your other rabbits might require. Their water and feed will need to be frequently checked in order to make sure that they have ample supplies of both at all times.

have one false dwarf, two true dwarfs, and one peanut. The false dwarf may or may not exhibit the qualities of breed type that you're seeking, so you may end up culling that one and selling it as a pet rabbit. The two true dwarf kits are quite nice and very typey, which is just what you were hoping for. And then there is the peanut.

Unfortunately, peanut kits are an inevitable fact of dwarf breeding, and the genetic combination of the double-dwarf gene is lethal. Although the kits are typically born alive, they die within a few days of birth. Their super-tiny size is a result of the double-dwarf genetics. Some breeders leave the peanut kits in the nest with the rest of the litter to allow nature to take its course, while others remove the peanut kits immediately and have them euthanized.

Is there any way to avoid producing peanut kits? Well, if you breed a false dwarf with a true dwarf, it is impossible to produce a peanut kit because the false dwarf does not possess a dwarf gene, and therefore the kits cannot inherit a dwarf gene from each parent. This successfully eliminates the

possibility of peanut kits. However, it also increases your odds of producing false dwarfs in your litters by making your odds 50 percent false dwarf and 50 percent true dwarf. Yet, if you consider that you're still producing an average of only 50 percent true dwarfs when crossing two true dwarfs, it may be an option worth considering, figuring that the additional 25 percent of false dwarfs are replacing the 25 percent of peanuts that you would be losing anyway. At least in this equation, you're producing an extra 25 percent of saleable pet-quality rabbits rather than losing 25 percent of your litter. However, it should be noted that you may find that your litters are not as consistent in type or quality if you are using false dwarfs in your breeding program.

CARING FOR KITS

The litter has arrived, they're happily tucked into their fur-lined nest box, you've done a quick cleaning of the nest box, and you're congratulating yourself on a healthy litter. But now you have a nest box full of responsibility, so what do you do next?

Don't underestimate the bond between a doe and her kits. Like any mother, a doe has a certain amount of protective instinct, as well as affection, toward her young.

For the first week or so, your doe will be the one handling the majority of the care for the litter. In these early days, the most important thing is for the doe and her litter to be left in peace and quiet with only the most minimal intrusions when necessary. You will want to do a daily check to make sure that each kit is still alive and to check that each one is getting fed. If the weather is particularly inclement (excessively hot or cold), you might want to consider bringing the nest box indoors to a temperature-controlled area for the majority of the day, returning it to the doe's hutch at least twice a day so that she can feed her kits. However, it's always better to leave the nest box in the doe's hutch if possible.

On the ninth day after the kits are born, it's time to do a thorough cleaning of the nest box itself. The easiest way to do this is to have a second nest box ready, freshly cleaned, and filled with hay. Remove the original nest box, transfer the babies to their new box, and promptly return the new box to the doe's hutch. The original nest box can be thoroughly emptied, disinfected, and prepared for a future litter.

If you don't have a second nest box available, carefully move the babies to a safe location while you empty the dirty nest box and remove all of the soiled hay and shavings. Refill the box with fresh bedding, place the babies back in the box, and slip it quietly back into the hutch.

You will want to keep a close watch over the kits' eyes, particularly as the 12th day approaches, when their eyes typically begin to open. If you notice that an eye seems to be "stuck," then you may want to slightly nudge it open; take care to be very cautious and gentle. Once the babies' eyes are opened, watch for any signs of eye infection, which are occasionally found in newborn kits. If one of your kits appears to have an eye infection, your veterinarian should be able to provide you with medication to treat this issue.

When the kits reach two to three weeks of age, they will begin to venture outside of the nest box for short periods of time. Don't be surprised if sudden or unexpected noises cause the entire litter to flee for the safety of the nest box; it's their safe haven and obviously the place to which they will run if at all frightened.

From this point until weaning time, the litter will continue to become more and more outgoing and engaging. They will begin nibbling on bits of hay or pellet pieces and sipping water, and of course growing like crazy.

WEANING

An inevitable part of raising a litter is weaning time. There are many opinions on the subject of weaning kits and the proper age at which to do so. Generally speaking, many rabbit owners prefer to wait until

TO CULL OR NOT TO CULL? THAT IS THE QUESTION.

The word "culling" is typically used to describe the process of eliminating undesirable examples from a breeding herd and focusing only on the very best rabbits that most closely represent the breeder's ideals. Culls, therefore, are the substandard rabbits that do not (for one reason or another) fulfill a breeder's expectations. What does a responsible breeder do with the culls?

There is a distinction between an animal that isn't suitable for your own breeding program and an animal that isn't suitable for anyone's breeding program. If you're planning to raise rabbits on any scale at all, the logistics are such that you're not going to need every rabbit that you produce for the future of your breeding program. Therefore, you may very well be selling perfectly good rabbits (i.e., not culls) to other breeders for use in their breeding programs, even though the rabbits don't fit your own needs.

However, if a certain rabbit is not suitable for your breeding program, is not suitable for anyone else's breeding program, and is unsuitable as a show specimen, then you could not in good conscience sell the rabbit to another breeder or show enthusiast as an acceptable example. So what are your other options?

For many people, culled rabbits are used for meat purposes. This has long been a common option for rabbits that are unsuitable for breeding or showing. For others, culled rabbits are sold as pet-quality rabbits. Many people (especially children) enjoy the company of a pet rabbit, and to them, it doesn't matter in the slightest if the rabbit is conformationally incorrect or has a disqualification that prevents it from competing in shows. Many breeding program culls have gone on to be beloved family pets.

Always remember that the decision to cull a rabbit from your herd is not something to make in haste. This is particularly true with litters of kits. Take your time, let the kits mature, and don't make any major decisions until the kits are at least eight weeks old. Prior to that age, it is extremely difficult to evaluate whether or not a kit will be a suitable breeding or show animal, especially if you are a novice breeder.

With regard to colors, it's somewhat easier to determine culls. For instance, if you are raising a breed that is only found in a very specific color or color pattern, it's easy to tell early on whether the kits in your litter exhibit the proper color or pattern. If your litter contains colors that are unrecognized for your breed by the ARBA, these rabbits will have to be culled from your program, unless you are attempting to work toward the development of a new recognized color for your breed (see more on this in Chapter 6) or your knowledge of color genetics is sufficient to allow you to use these unrecognized colors to produce recognized colors.

the kits are eight weeks old before weaning them, although some people will wean at six weeks and others as early as four weeks. Some of the timing decision has to do with whether the doe has been bred back for another litter, and some has to do with how many litters a doe is going to produce each year. If you are aiming for more than four litters from each doe per year, you will have to wean your kits at no later than eight weeks in order to achieve this goal. Some commercial rabbitries will push for five or six litters per year, which results in earlier weaning dates for the doe's litters. If maximum production is not your aim, you may want to take things a bit slower and allow your kits to nurse for up to 10 weeks before gradually weaning them.

Weaning does not have to take place all at once. Your own instincts will help guide you in the best course of action for your does and each individual litter. Let's say you have a litter of five kits. After the eighth week, you might want to wean the largest three kits and leave the smaller two with the doe for a bit longer to give them a bit of an added boost. Another consideration with regard to weaning is that if your kits appear to be overweight at around six or seven weeks, you may want to think about weaning them a bit earlier to prevent them from continuing to gain weight.

How about the newly weaned kits? These growing youngsters will be eating plenty, yet you want to be certain that they are getting the right amounts to support their growth. At the same time, you want to do your best to protect them from enteritis, a dangerous cause of diarrhea in weanling rabbits.

Before you begin weaning, it's wise to take some time to observe your litter. Are they eating pellets well? Are they munching on hay? Are they regularly drinking from the water bottle or from the water crock? These are all good signs that your litter is sufficiently ready for weaning. After you move them from the doe's hutch into their own, you will want to be vigilant about providing constant access to clean, fresh water. Remember, these kits do not have the benefit of hydration through nursing any longer, so water is the necessary replacement. Secondly, you will want to provide access to quality grass hay. This is very important. Hay provides a good source of the fiber that is necessary to a young rabbit's healthy digestion. Some breeders underestimate the importance of hay in a rabbit's diet, but it is a vital component, and it's especially important for young rabbits to have access to hay. Pellets are also fed to young rabbits, but pay close attention to the amount of protein in the pellets you purchase. Many breeders prefer to feed pellets that contain 16 percent protein or less to weanling rabbits, while other breeders prefer an 18-percent-protein pellet. Generally speaking, growing kits should be offered as much hay and pellets as they can eat, but not more than they will eat. If you're feeding too much, you'll find leftover pellets in the feeder or trampled hay in the cage, which is unnecessary waste.

Young rabbits enjoy attention, so take advantage of this natural-born curiosity and take the time to frequently handle your young kits.

It's important to pay special attention to your newly weaned kits. Always make sure that each one is eating and drinking well, as they will no longer be receiving nourishment from their mother's milk.

Baby rabbits may jump at the chance to munch on fresh greens and vegetables, but many breeders are hesitant to offer theses types of food for fear of enteritis.

GREEN ACRES FOR KITS?

Of all of the topics that are perpetually debated among rabbit raisers, this is one for which there are always multiple opinions: should young rabbits be allowed access to fresh green food or not? Each side of the debate has valid points, and it's certainly up to each rabbit breeder to evaluate their own opinion and feelings on the subject.

In the wild, baby rabbits begin eating green food right from the start by munching on grass blades, dandelion leaves, and bits of clover. Obviously, in the controlled environment of a hutch, you are the one providing the food, and you certainly aren't required to offer an identical diet to that of a wild rabbit. So should you offer access to fresh greens or not?

Some breeders feel strongly that you should not feed greens to baby rabbits because young rabbits are at risk of mucoid enteritis, which is a cause of diarrhea and sometimes death. Their immature digestive systems are still developing, and some breeders feel that there is a connection between the feeding of greens and subsequent incidence of mucoid enteritis.

On the other hand, some breeders feed greens to all of their rabbits (including the babies) every day without any ill effects. These breeders feel that since a rabbit's natural diet includes such delicacies as grass, dandelion leaves, and other fresh greens, it only makes sense that these types of food are safe for consumption by weanling rabbits. The important factor to remember is moderation. You wouldn't want to take an eight-week-old kit, wean him from his mother, put him in his own cage, and give him an enormous stack of dandelion leaves. This is only common sense. Any rabbit's digestive system needs time to adjust to a change in diet, whether it's from one brand of feed to another, or from grass hay to alfalfa, or from not eating greens to eating them. If you choose to feed greens to your weanling kits, proceed slowly. Feed only a small amount the first day and slowly increase the amount over the course of a week until you've reached the amount that you desire to feed each day.

CHAPTER 10
SHOWING RABBITS

If you're just starting out with rabbits, it probably won't be long before you are intrigued with the idea of attending a show. You may decide to start out with something small like your local county fair, but it's probably inevitable that sooner or later you're going to want to try showing at an ARBA-sanctioned show. And why not? Rabbit shows bring together dedicated enthusiasts, allow spectators to view a multitude of breeds in one location, give breeders the opportunity to showcase their stock, and provide an educational experience for exhibitors and spectators alike. Besides all of this, rabbit shows are a wonderful place for buyers and sellers to network and make new acquaintances. It's no wonder that you might want to join in the fun!

Here is a happy sight for any rabbit enthusiast. Rabbit carriers filled with beautiful show specimens are a common sight at any rabbit show.

Holland Lops usually have some of the highest entry numbers at ARBA shows, and competition is stiff. If competing in large classes appeals to you, you'll have a great time showing Holland Lops.

It has been estimated that approximately 850,000 rabbits are shown each year across the United States at around 3,000 ARBA-sanctioned shows. Chances are good that you'll be able to locate a show (or shows) in your area, either to attend as a spectator or to exhibit your bunnies. Although it's possible that you might be driving down the road one day and happen across a sign that says "Rabbit Show Today," chances are that it might take a bit of detective work in order to find the shows in your area.

The best place to start is the ARBA website or the ARBA's official magazine, *Domestic Rabbits*. Both of these resources list upcoming shows, categorized by state, so that you can quickly locate the shows that are close to you. Another great way to keep on top of upcoming shows is to join any local or regional rabbit clubs. This will guarantee that you are kept informed of the latest show plans and dates for shows within your area. You can also find out if your ARBA district runs an Internet Yahoo group for discussion of events and news within your district. This is another great resource for staying abreast of upcoming shows. Your breed's specialty club is also a good place to find information on upcoming shows, including your breed's national show. Once you've found a show to attend, it's time to select which rabbits you're going to show.

CHOOSING RABBITS TO SHOW

It takes experience, knowledge, and a bit of finesse to choose rabbits for showing purposes. Although choosing a rabbit from your barn could be as simple as closing your eyes and going "eenie, meenie, minie, moe," you will undoubtedly have better success on the show table if you carefully select only your best specimens to show.

It's important to fully understand the various divisions that are offered at ARBA shows and to always double-check your entry forms to be sure that your rabbits are properly entered.

WHAT'S A 6/8?

The basic divisions at rabbit shows are Senior Buck, Senior Doe, Junior Buck, and Junior Doe. However, in any breed that matures to a standard senior weight of over 9 pounds, there is an additional show division, called the 6/8s, or the Intermediate division. The Intermediate classes are for rabbits between the ages of six and eight months—the "intermediate" aged rabbits—and it allows them to participate in a class specifically suited to their age without having to compete against fully mature rabbits.

Let's say, for instance, that you're showing a seven-month-old Siamese Satin buck. Your first class would be for 6/8 Siamese Satin bucks. If you were fortunate enough to win this class, your rabbit would go on to compete against Satin bucks from all the other colors in the 6/8 age group. If your rabbit were to prevail again in this class, he would go up against the junior buck, junior doe, 6/8 doe, senior buck, and senior doe winners to compete for the Best of Breed championship.

Most breeds have senior weights that are under 9 pounds, so the 6/8 classes are only for a handful of breeds: American, Giant Angora, Beveren, Californian, Champagne d'Argent, Checkered Giant, American Chinchilla, Giant Chinchilla, Cinnamon, Creme d'Argent, Flemish Giant, Hotot, English Lop, French Lop, New Zealand, Palomino, and Satin.

It goes without saying that breed type is one of the most important factors to consider when choosing a rabbit to show. The rabbit must exemplify its breed's entry in the *Standard of Perfection* in order to place well on the show table, so choosing your show rabbits based on type is always an important first step.

The age of your rabbit is another important consideration. Many breeders feel that a rabbit is past its prime show condition by the age of two years old, so showing a rabbit that is older than two may result in lower placings than the rabbit may have previously won. This is especially true in the

Your rabbits must meet the weight requirements for the class in which you've entered them, so always weigh them prior to heading off to a show.

When you are selecting rabbits to show, it's wise to compare each rabbit's characteristics in order to choose the rabbits that most closely resemble the description in the ARBA's *Standard of Perfection*.

It's fascinating to observe an ARBA judge at work. Their comments and evaluations of each rabbit can be a very insightful and educational experience.

Rabbit shows are a great place to gather information about local and state rabbit clubs. Membership literature and pamphlets are usually available at rabbit shows.

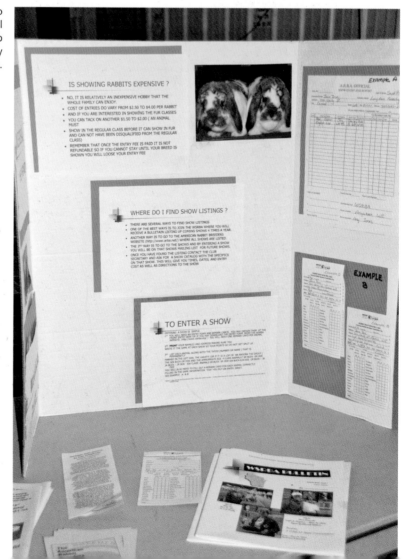

These show rabbits are happily situated in their show carriers, complete with water dishes, food dishes, and hay. Check your rabbits frequently during the show to make sure that they always have fresh water and food available.

THREE LEGS AND COUNTING

One of the strangest components of rabbit lingo is undoubtedly the use of the word "leg" to signify specific wins at ARBA shows. It can be a little unsettling when a person is initiated into the world of rabbits only to discover that some rabbits have 3, 5, or 14 legs!

The term "leg" is used to describe any of several situations. If a rabbit places first in a class that consisted of five or more rabbits entered by three or more exhibitors, the winning rabbit is awarded a leg. By the same token, winning a category such as Best of Group or Best of Breed (with the same qualifications as listed above) also results in a leg. There are a few other rules that pertain to the awarding of legs, but as long as all of the criteria are met, when a rabbit obtains three legs, the owner can apply for a Grand Champion Certificate from the ARBA.

It can be difficult to obtain legs for your rabbits in the best of times, but this is particularly true if you are showing one of the rare breeds. Because of the lack of entries in classes for rare breeds, you may not have enough competition to qualify your rabbit for a leg, even if he wins his class or a Best of Breed award. Therefore, the process of obtaining three legs and achieving a Grand Championship can take a lot longer when you're showing in classes that have few entries.

It's very rewarding to have the quality of your breeding program validated by an impressive show win, and the trophies are an added bonus.

case of does that have produced litters, as kindling can affect their overall condition and leave them looking a bit "matronly."

Some rabbit breeders choose to carefully calculate their breedings so that they result in litters that will be the optimal age for showing by the date of an important show. For instance, if your breed's national show is held in July and you're planning to show juniors there, you will want to aim for juniors that are as close as possible to six months old on the date of the show. It is more advantageous to show juniors that are as mature-looking as possible. Therefore, you may want to calculate your breeding dates in order to produce youngsters that will be nearly six months old (but not over) by the date of the show in July.

The current condition of your show rabbits is also something to consider. A rabbit that is going through the process of molting is not in any condition to be shown, so if your favorite show rabbit chooses to molt around the date of your upcoming show, then it's time to select another rabbit and leave the molting one at home.

Weight is another aspect that is very important when choosing show specimens. Your rabbit's weight must fall within the standard for its breed, so always take a moment to weigh it before you head out to a show. You might be surprised to discover that your big buck is just a tad bigger than he should be.

Good health is an area in which there can be no compromise. You should never show a rabbit that is anything but the picture of perfect health. Besides being an area for disqualification, you would not want to expose the other rabbits at the show to any type of illness, nor would you want to add any stress to your own rabbit while he or she is trying to recover from any type of illness or injury.

It's no fun to go to all of the trouble of attending a show only to have one of your rabbits disqualified (DQ'd) from competition upon arrival. Therefore, it's always advisable to carefully go over the *Standard of Perfection* and familiarize yourself with all of the different types of DQs in order to determine whether your rabbit exhibits any of these faults. Carefully check toenails before you show; the same goes for dewclaws and teeth.

CONDITIONING

As with so many areas of life, showing rabbits takes patience. The process of fully conditioning your rabbits for showing does not happen overnight. A rabbit in prime show condition has weeks or months of preparation behind it, so it's wise to be thinking about your show prospects far in advance of the show date.

Good conditioning starts with the basics: fresh water and quality food. As we've already discussed in Chapter 4, it is essential to give careful consideration to your feeding program, and this is especially true in the case of conditioning rabbits for show.

But what about supplements? This is an area in which most rabbit enthusiasts have developed

Supplements are often used to increase a rabbit's condition for show. Some breeders prefer natural supplements, such as oats, barley, or sunflower seeds, while other breeders use professional conditioning products, such as Show Bloom, seen here.

There are hundreds and sometimes thousands of rabbits at a single show, so it's very important that you only exhibit rabbits that are completely healthy.

Proper posing is a very important aspect of showing. Study your *Standard of Perfection* to understand the intricacies of the proper pose for your breed.

their own methods that work best for them, so you may want to carefully experiment in order to find out which supplements are the most beneficial in your own show conditioning program. Some ideas to try are barley, oats, wheat germ oil, and sunflower seeds.

Many breeders feel that the conditioning of rabbits is somewhat influenced by genetics, so if you are having difficulty in producing rabbits that possess good overall condition (firm flesh and excellent fur), you may want to reconsider the breeding stock that you are working with. By selecting to breed only from rabbits that maintain good general condition, you will help to reduce the number of poorly conditioned rabbits in your barn.

In the case of the wooly breeds (the Angoras, the Jersey Wooly, and the American Fuzzy Lop), it takes dedicated daily grooming and preparation in order to be successful on the show table. Because of the emphasis on wool in these breeds, the condition of your rabbit's fur is especially important, and your

rabbit cannot reach optimum condition unless you are a dedicated groomer.

Fur condition is important in all breeds, though, so you will want to pay close attention to the condition of the fur on your show rabbits even if you're not raising a wooly breed. On each breed's "schedule of points" (the basis upon which each breed is judged), fur is allotted a particular value of points. This can range from as low as 5 points (English Lop and Rhinelander) to as high as 60 points (the Satin Angora), so fur is an important characteristic and one that you will definitely want to pay attention to, no matter what breed you're showing.

POSING

As important as conditioning and breed type are on the show table, posing is another aspect that carries a lot of importance if you're serious about showing. A rabbit that cannot (or will not) pose properly will not be able to compete well. Even though it may sound complicated to the novice exhibitor, posing your rabbit properly and teaching him or her to pose properly are not difficult at all. It just takes a little education, a little practice, and a consistent approach.

All of the breeds in the Compact and Commercial types are posed in basically the same way, as described in the ARBA's *Standard of Perfection*. All of the Semi-Arch breeds are posed in the same manner, with the exception that some breeds may be moved naturally for further evaluation by the judge.

The Full-Arch breeds, which include the Belgian Hare, the Britannia Petite, the Checkered Giant, the English Spot, the Rhinelander, and the Tan, are given the opportunity to move about, although the Britannia Petite is posed.

The Himalayan is in a class by itself. As the only breed in the Cylindrical shape, the Himalayan is posed entirely differently than any other breed.

It is important to note a couple of variations on the breeds in the Compact group. Holland Lops and Netherland Dwarfs are posed somewhat differently than the other breeds in the Compact category. Holland Lops are not posed with their forelegs laying flat on the table, but instead in a more upright position that shows their chest and front legs very clearly. Netherland Dwarfs are posed in a more natural stance.

Knowing the standard pose for your breed is very important. You wouldn't want to teach your Britannia Petite to pose like a Polish and then take her to her first show only to be disappointed. Always refer to the *Standard of Perfection* in order to determine the proper method of posing for your particular breed.

NETWORKING

Let's imagine that you specialize in the Blue variety of Dutch rabbits. You show your rabbits frequently and they usually do very well on the table. However, because you operate a small rabbitry, you raise only five or six litters per year and only occasionally have animals for sale.

Sometimes a rabbit just catches your fancy. Imagine coming upon this delightful Tortoise Dutch junior doe for sale at a show. Despite any initial infatuation, always thoroughly examine the rabbit for any faults or DQs prior to purchase.

PAPERWORK AND DETAILS

If this is your first ARBA show, before you enter the competition you will definitely want to refer to your ARBA Official Guide Book, which is the complementary book you received when you joined the ARBA. The guide book contains an excellent article on "Entering a Sanctioned Show" that fully details the process of filling out the show entry blank. It also discusses several commonly asked questions about ARBA-sanctioned shows; it fully explains many topics, such as age at the time of the show and tattooing; and includes several helpful hints. You will feel a lot more confident about heading out to a show after you've reviewed this helpful information.

Whenever you're preparing for a rabbit show, it's always wise to double-check all details before you hit the road. Even if you've attended many shows in the past, there's always the chance that you might overlook or forget something, so take a few minutes to go over the important details.

Check tattoos: Your rabbit cannot be shown without a clearly legible tattoo in its left ear. This is vitally important, so always double-check that your rabbit is tattooed and that the tattoo is clearly readable. Never overlook this detail.

Check toenails and teeth: Toenails that are improperly colored can be a cause for disqualification, so always double-check that your rabbit's toenails are of the appropriate color before you head out to a show. The same goes for the rabbit's teeth. Take a quick peek into your rabbit's mouth ahead of time so that there are no surprises at the show.

Check your classes: Always make sure that your rabbit is entered in the proper class. This means that you should double-check your rabbit's age to make sure that you aren't inadvertently entering a junior in a senior class (or vice versa). Although it seems like an unlikely mistake, you will want to double-check that the rabbit you're entering is the proper sex for its class. You will be disqualified for accidentally entering a buck in a doe's class, so always double-check before heading out to the show.

Check your supplies and equipment: It's no fun to arrive at a show only to discover that you've left something important behind. Hopefully it isn't your rabbit. It's inconvenient to leave behind something that you really need, such as hay or feed or water. A pre-show checklist can help to keep track of all of the necessary items that you need for showing and may help prevent you from driving off without something that you really and truly need.

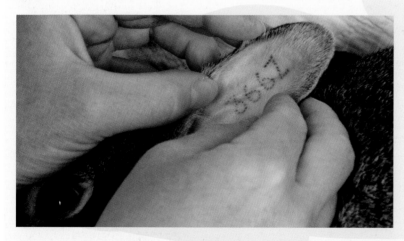

Double-checking your rabbit's tattoo is one of several details that you will want to verify before heading off to a show.

Rabbit owners often set up along the floor at a rabbit show and surround their personal grooming table with their various carriers and supplies.

At the rabbit shows, you often compete against another breeder (we'll call her Karen) who also specializes in the Blue variety of Dutch. You have a high regard for the quality of Karen's rabbits, and she feels the same way about yours.

Let's say that you receive an inquiry from someone who is looking for a pair of Blue Dutch to begin a breeding program. You've just sold the last one from your most recent litter and have nothing available for sale, so you mention Karen's rabbits and the fact that she might have some available.

A couple of months later, you have some new litters, and one day you receive another inquiry from someone who says, "I am looking to purchase a couple of Blue Dutch, and Karen Jones recommended that I contact you. She really praised the quality of your rabbits and I am wondering if you have any available for sale at this time?" That is the essence of networking. By helping others, you receive help in return.

Networking "works" in other ways, too. By networking, you can trade rabbits with other breeders, arrange transportation for purchased or sold rabbits, brainstorm new ways to increase the interest in your breed, carpool to shows, and help each other to make sales. All of these are excellent ways for breeders to work together toward the greater goal of enjoying your rabbits to the fullest extent.

Catalog shopping is all well and good, but seeing the products in person can make your selections much easier. This rabbit equipment supplier has set up wares at a rabbit show.

Rabbit carriers have flip-top lids that are fastened with a spring and a latch. This makes for easy access to your rabbit, but is also secure enough to prevent him or her from escaping.

TAKE ME TO THE SHOW

Cage Tags, Inc.
www.cagetags.com

RABBIT SHOW REMARK CARD
American Rabbit Breeders Association, Inc.

Ear No. _____ Coop No. _____ Entry No. _____ X
Exhibitor _____ X
Address _____ X
Show _____ Date _____ X
Breed _____ Variety _____ X

| Buck | Doe | Sr. | 6/8 | Jr. | Pre Jr. | Fryer | Meat Pen | Fur | X |

No. in Class _____ Award _____ No. Exhibitors _____
B.O.B.   B.O.S.   B.O.G.   B.O.S.G.   B.O.V.   B.O.S.V.
Best Sr.   Best 6/8   Best Jr.   Best Pre-Jr.

	VG	G	F	P			VG	G	F	P
Head					Condition					
Ears					Butterfly					
Crown					Eye Circles					
Bone					Cheek Spots					
Type					Ear Base					
Shoulders					Side Markings					
Midsection					Spine/Herringbone					
Hindquarters					Blaze					
Fur/Wool					Cheeks					
Sheen					Neck					
Density					Saddle					
Texture					Undercut					
Color					Stops					

Remarks _____

It's easy to get confused over which rabbits you've brought to show and which rabbits you've brought to sell, so make things easy on yourself by clearly marking your show rabbits.

A judge's remark card can help you to better understand how and why your rabbit placed the way it did at a show. These cards are very helpful and beneficial, especially when you're just starting out.

When marking your carriers that contain the rabbits you'll be selling at a show, you might consider placing a pink "for sale" tag on the carriers with does and a blue "for sale" tag on the carriers that contain bucks.

FOR SALE

Cage Tags, Inc.
www.cagetags.com

FOR SALE

Cage Tags, Inc.
www.cagetags.com

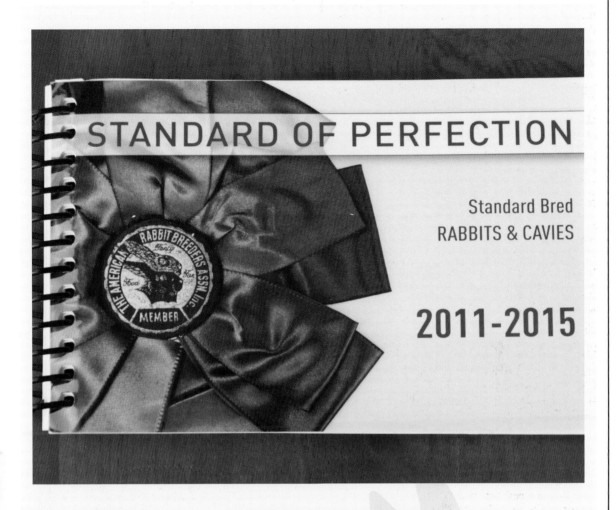

THE *STANDARD OF PERFECTION* IS YOUR FRIEND

Anyone who wishes to show rabbits needs to be in possession of the ARBA's *Standard of Perfection*. It is a wonderful spiral-bound book filled with the breed standards for each and every one of the 47 breeds recognized by the ARBA. For each breed, the book includes the standard weights, the recognized colors and varieties, the schedule of points (for judging), disqualifications, and faults (and the differences between the two; the distinction is important to recognize), and a photo of the breed. In addition to being very important for anyone interested in showing, it's also a book that anyone raising rabbits will want to have on hand, as you must be intimately familiar with your breed's standard if you intend to raise quality stock that meets the criteria for the breed.

The *Standard of Perfection* is revised every five years, and if you want to keep up-to-date on the current breeds, varieties, and changes, then you will want to purchase each new edition as it is released. This is a book that every rabbit owner should have in his or her home library, and it's one that you will refer to again and again. It is available through the ARBA (www.arba.net) or also through rabbit supply catalogs.

It might be said that networking is just a fluffed-up term for friendship, but it means more than that as well. Networking involves a certain awareness of "who needs help" and "who can help you." For instance, you may be in the market for a top-quality Thrianta buck. You would like one that is a splendid show specimen and could also make a good foundation for your future expansion into breeding Thriantas. Now, you can tell your best friend, a Mini Rex aficionado, that you're on the lookout for a Thrianta buck, but this may or may not result in any leads toward finding this particular animal. On the other hand, you can network with a fellow show exhibitor that raises English Angoras and also has two Thriantas. She can check among her acquaintances and would have a far higher likelihood of helping you to locate a reputable Thrianta breeder with quality stock.

It's important to remember that networking, like so many other things in life, works both ways. As wonderful as it can be to have assistance and help in achieving your own goals and needs, it's equally important to be helpful to others and assist when you can be of service. Perhaps you can deliver a friend's rabbit on your way to a show, or perhaps you can point someone in the direction of a breeder who might have just the rabbit that they have been searching for. If you help others, they will be much more likely to help you, and that's a definite win-win situation for everyone.

MARKETING

Rabbit shows are one of the best marketing venues that a breeder could ask for. As we'll discuss in a moment, showing helps to validate the quality and reputation of your stock, which works to increase

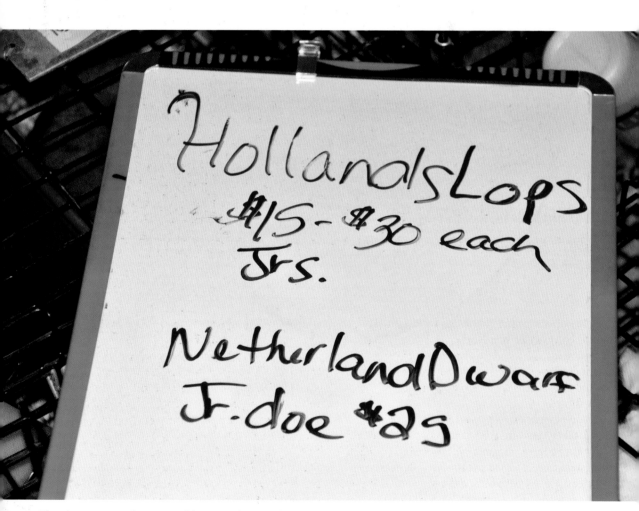

The chance to market your rabbits at a show is always an added bonus of attending. Marker boards make it easy to prepare a sign that advertises what you have available for sale.

This judge is comparing two rabbits in order to determine the winner of the class that she is judging. Rabbits are judged in comparison to the breed's description in the *Standard of Perfection*, and also are judged in comparison to each other.

the value of your rabbits. You can (and probably should) consider print advertising and a website, but you really can't underestimate the importance of actually having your rabbits out in the public eye, and that is most easily accomplished at rabbit shows.

If you have rabbits that you're interested in selling, take them along to a show and let the world see what you have available. These could include youngsters from your recent litters (as long as they are at least eight weeks old), older stock that you're no longer using or no longer have a need for, or individuals of a certain color that you're phasing out of your program. If you want to market your rabbits

quickly and easily, taking them along to a rabbit show is probably your best bet.

One caveat is that rabbit shows are probably not the ideal place for you to market your pet-quality rabbits. This would include rabbits that you've culled for improper type, rabbits that are mismarked, and rabbits that are outside the proper weight range for their breed. This is not to say that you won't find buyers for your pet-quality rabbits at a show, but your potential customers are more likely to be searching for a show-quality (or at least brood-quality) rabbit. You are probably better off trying to market your pet-quality rabbits through local venues, such as pet shops, newspaper ads, and pet swaps.

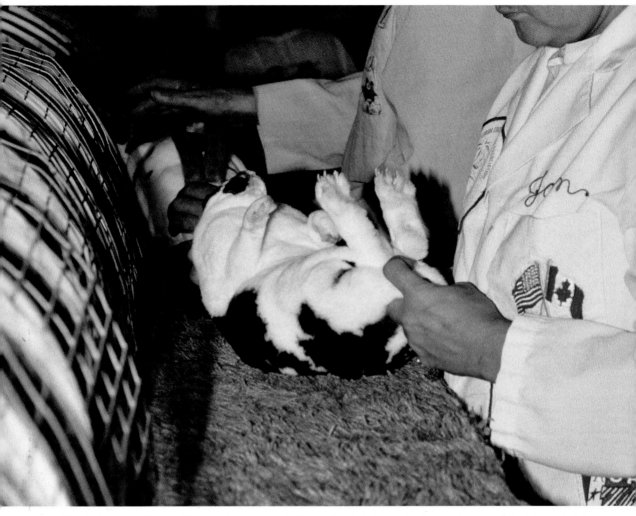

A rabbit judge must thoroughly examine every rabbit on all sides—front, back, top, bottom, left side, and right side—in order to fully evaluate the animal.

Being a regular exhibitor at shows also helps to keep your rabbitry fresh in people's minds. If you're exhibiting your Blue Havanas at every single show within a 300-mile radius of your home, sooner or later your name (and your rabbitry's name) will be connected with Blue Havanas in people's minds. You would hope to have the name of your rabbitry eventually become synonymous with Blue Havanas in general and quality Blue Havanas in particular, so that when a buyer is looking for quality Blue Havanas, the first name that everyone recommends is yours. Obviously, this is not something that takes place overnight; establishing yourself and your rabbitry is a lengthy and time-consuming process. However, consistent showing and consistently high placings at shows reap dividends later on.

PROVING YOUR STOCK

You believe in your breeding stock. You have to; otherwise you wouldn't be breeding them. That's not to say that there aren't areas in which you might like to improve certain characteristics, but on the whole, you have to believe in the quality of your stock or else you would have culled them long ago and moved on to something else. Still, even if you believe that your rabbits are of above-average quality, it may take time to convince your potential clients that your rabbits are something extra-special.

One way to do this is to get out and show. By showing your stock at ARBA-sanctioned shows, you are putting your rabbits up against other rabbits of the same breed, variety, and age, which makes a level playing field when it comes to determining where your rabbits fit into the grand scheme of the breed. If your rabbits are consistently placing well and holding their own on the show table (or better yet, if they are winning and achieving Best of Breed awards), this will go a long way toward proving the quality of your stock to potential customers. It's easy to sing the praises of your rabbits and tell everyone that they are absolutely wonderful, but it really helps to have something concrete with which to back up your assertions. Show records will accomplish this. If you can say that your senior Opal Holland Lop has three legs, the statement carries much more weight than saying that your senior Opal Holland Lop is "absolutely fabulous and a really great rabbit." Show records help prove the quality of your stock.

By showing your foundation stock and validating the quality and type you're working with, you can command a higher price for your young rabbits than you could otherwise ask if your foundation stock was unproven.

In addition to proving your stock in the eyes of others, it's possible that you may very well want to prove your stock in your own eyes. When newcomers begin rabbit raising, it can often be difficult to truly determine whether or not their breeding stock is "up to fluff." The nuances and intricacies of the breed's description in the *Standard of Perfection* can be hard to tangibly connect to the rabbit in your hands. Therefore, newcomers often like to show their foundation stock in order to get a judge's expert opinion on the quality and type of their rabbits. In a way, shows can help to prove your stock not only for your clients, but also for yourself.

FRIENDSHIP AND FUN

Who doesn't need a little fun in their life? And if you can have fun at the same time that you're promoting your rabbitry, how much better can it get? Many rabbit enthusiasts attend shows for the sheer joy of it. The thrill of showing, the good-natured competition against other exhibitors with quality rabbits, the delight of seeing a multitude of breeds in one location, the camaraderie, and the chance to travel and see new locations. It's a wonderful combination, especially when coupled with the other benefits of proving your stock, networking, and marketing your rabbits.

At a show, each rabbit is temporarily placed in a small cage on the judging table. This allows the judge to quickly access each entry to evaluate and compare them

RESOURCES

RABBIT EQUIPMENT SUPPLIERS

Bass Equipment Company
www.bassequipment.com
800-798-0150

BunnyRabbit.com
www.bunnyrabbit.com
bunnies@bunnyrabbit.com

KD Cage Company and Supplies
www.kdcage.net
800-265-5113

Klubertanz Equipment Company, Inc.
www.klubertanz.com
800-237-3899

Koenig's Kountry Supplies
www.koenigskountrysupplies.com
888-655-0351

KW Cages
www.kwcages.com
800-447-CAGE

BOOKS, PERIODICALS, AND ORGANIZATIONS

American Rabbit Breeder's Association (ARBA)
P. O. Box 5667
Bloomington, IL 61702
www.arba.net
If we can give you one piece of advice, it would be to join the ARBA. You will receive six issues of Domestic Rabbits *magazine per year, as well as the ARBA yearbook and their* Official Guide Book to Raising Better Rabbits and Cavies.

Rabbits USA
www.SmallAnimalChannel.com
An annual magazine jam-packed with articles and photos relating to rabbit care.

The Field Guide to Rabbits, by Samantha Johnson
www.voyageurpress.com
A photographic guide to the 47 rabbit breeds recognized by the ARBA. The book also includes information on shapes, sizes, colors, and fur varieties. An excellent companion to How to Raise Rabbits.

The Rabbit Book, by Samantha Johnson
www.voyageurpress.com
An excellent guidebook for young rabbit enthusiasts. Explains everything you need to know about housing, feeding, showing, and care.

RABBIT BREED CLUB WEBSITES

American: www.americanrabbits.org
American Chinchilla:
http://americanchinchillarabbitbreedersassociation.com/
American Fuzzy Lop: http://aflrc.weebly.com/
American Sable: www.americansables.webs.com/
Angoras (English, French, Giant, and Satin):
http://nationalangorarabbitbreeders.com/
Belgian Hare: www.belgianhareclub.com/
Beveren: www.freewebs.com/beverens/index.htm
Blanc de Hotot: www.hrbi.org/
Britannia Petite: www.britanniapetites.com/
Californian: www.californianrabbitspecialtyclub.com/
Checkered Giant: www.acgrc.com/
Crème d'Argent: www.cremedargent.org/
Dutch: www.dutchrabbit.com/
Dwarf Hotot: www.adhrc.com/
English and French Lops: www.lrca.us/
English Spot: http://americanenglishspot.webs.com/
Flemish Giant: www.nffgrb.com/
Florida White: www.fwrba.net/
Giant Chinchilla: www.giantchinchillarabbit.com/
Harlequin: www.americanharlequinrabbitclub.net/
Havana: www.havanarb.org/
Himalayan: www.himalayanrabbit.com/
Holland Lop: www.hlrsc.com/
Jersey Wooly: www.njwrc.net/
Lilac: www.nlrca.webs.com/
Mini Lop: www.amlrc.com/
Mini Rex: www.nmrrc.net/
Netherland Dwarf: www.andrc.com/
New Zealand: www.newzealandrabbitclub.net/
Palomino: www.palominorabbit.com/
Polish: www.americanpolishrabbitclub.com/
Rex: http://nationalrexrc.org/
Rhinelander: www.rhinelanderrabbits.com/main.htm
Satin and Mini Satin: www.asrba.org/
Silver: www.silverrabbitclub.com/
Silver Fox: www.nsfrc.com/
Silver Marten: www.silvermarten.com/
Standard Chinchilla: www.ascrba.com/
Tan: www.atrsc.org/
Thrianta: www.atrba.net/

INDEX

ABOUT THE AUTHORS

Daniel Johnson, writer and photographer, likes to spend his time lugging around heavy camera equipment in all kinds of weather to take pictures of things like dogs pulling sleds at -20 below or people hauling hay at 90 above. He loves to photograph horses as well, and he is the co-author (with Samantha) and photographer of *Horse Breeds: 65 Horse, Draft, and Pony Breeds*; *How to Raise Horses: Everything You Need to Know*; and the Horse-a-Day box calendar, all from Voyageur Press. He's also the author and photographer of *The 4-H Guide to Digital Photography*. During the making of *The Beginner's Guide to Beekeeping*, Dan discovered that it's possible—though not easy—to take bee pictures from behind a beekeeping veil. In his spare time, he also photographs frogs, one of which has been his pet for the last twenty years.

Samantha Johnson is an award-winning writer, as well as a proofreader and pony wrangler. On any given day, you might find Samantha pursuing a variety of occupations: crafting words into articles, advertisements, or books; planting heirloom vegetables in the garden; or harvesting honey from the hives. On another day, she might be hauling hay, feeding livestock, or assisting with the delivery of a newborn foal. She is also a horse show judge, certified with the Wisconsin State Horse Council and the Welsh Pony and Cob Society of America, and has judged horse shows across the United States from Maryland to California and locations in between. Samantha is the author of several books, including *The Rabbit Book*, *The Field Guide to Rabbits*, and the co-author (with Dan) of *The Beginner's Guide to Vegetable Gardening*. Samantha enjoys making to-do lists, watching old episodes of *Little House on the Prairie*, and daydreaming about buying a couple of Cheviot sheep and a Miniature Jersey cow.

As brother and sister collaborators, Dan and Samantha pursue their writing, photography, and agricultural interests at the family-owned Fox Hill and Pine Valley Farms in northern Wisconsin. Since 1999, they have been involved with raising and showing registered Welsh Mountain Ponies, and they also keep an assortment of purebred rabbits, including Mini Rexes and Holland Lops. Several hundred thousand honey bees also make their home at Fox Hill and Pine Valley, which keeps life sweet.